## BIOTECHNOLOGY IN AGRICULTURE SERIES

**General Editor: Gabrielle J. Persley, Chair, The Doyle Foundation, Glasgow, Scotland.**

For a number of years, biotechnology has held out the prospect for major advances in agricultural production, but only recently have the results of this new revolution started to reach application in the field. The potential for further rapid developments is, however, immense.

The aim of this book series is to review advances and current knowledge in key areas of biotechnology as applied to crop and animal production, forestry and food science. Some titles focus on individual crop species, others on specific goals such as plant protection or animal health, with yet others addressing particular methodologies such as tissue culture, transformation or immunoassay. In some cases, relevant molecular and cell biology and genetics are also covered. Issues of relevance to both industrialized and developing countries are addressed and social, economic and legal implications are also considered. Most titles are written for research workers in the biological sciences and agriculture, but some are also useful as textbooks for senior-level students in these disciplines.

**BIOTECHNOLOGY IN AGRICULTURE SERIES**

---

**Titles Available:**

1: Beyond Mendel's Garden: Biotechnology in the Service of World Agriculture *
*G.J. Persley*
2: Agricultural Biotechnology: Opportunities for International Development
*Edited by G.J. Persley*
3: The Molecular and Cellular Biology of the Potato *
*Edited by M.E. Vayda and W.D. Park*
4: Advanced Methods in Plant Breeding and Biotechnology
*Edited by D.R. Murray*
5: Barley: Genetics, Biochemistry, Molecular Biology and Biotechnology
*Edited by P.R. Shewry*
6: Rice Biotechnology
*Edited by G.S. Khush and G.H. Toenniessen*
7: Plant Genetic Manipulation for Crop Protection *
*Edited by A. Gatehouse, V. Hilder and D. Boulter*
8: Biotechnology of Perennial Fruit Crops
*Edited by F.A. Hammerschlag and R.E. Litz*
9: Bioconversion of Forest and Agricultural Plant Residues
*Edited by J.N. Saddler*
10: Peas: Genetics, Molecular Biology and Biotechnology
*Edited by R. Casey and D.R. Davies*
11: Laboratory Production of Cattle Embryos
*I. Gordon*
12: The Molecular and Cellular Biology of the Potato, 2nd edn
*Edited by W.R. Belknap, M.E. Vayda and W.D. Park*
13: New Diagnostics in Crop Sciences
*Edited by J.H. Skerritt and R. Appels*
14: Soybean: Genetics, Molecular Biology and Biotechnology
*Edited by D.P.S. Verma and R.C. Shoemaker*
15: Biotechnology and Integrated Pest Management
*Edited by G.J. Persley*
16: Biotechnology of Ornamental Plants
*Edited by R.L. Geneve, J.E. Preece and S.A. Merkle*
17: Biotechnology and the Improvement of Forage Legumes
*Edited by B.D. McKersie and D.C.W. Brown*
18: Milk Composition, Production and Biotechnology
*Edited by R.A.S. Welch, D.J.W. Burns, S.R. Davis, A.I. Popay and C.G. Prosser*
19: Biotechnology and Plant Genetic Resources: Conservation and Use
*Edited by J.A. Callow, B.V. Ford-Lloyd and H.J. Newbury*
20: Intellectual Property Rights in Agricultural Biotechnology
*Edited by F.H. Erbisch and K.M. Maredia*
21: Agricultural Biotechnology in International Development
*Edited by C. Ives and B. Bedford*
22: The Exploitation of Plant Genetic Information: Political Strategies in Crop Development
*R. Pistorius and J. van Wijk*
23: Managing Agricultural Biotechnology: Addressing Research Program Needs and Policy Implications
*Edited by J.I. Cohen*
24: The Biotechnology Revolution in Global Agriculture: Innovation, Invention and Investment in the Canola Industry
*P.W.B. Phillips and G.G. Khachatourians*
25: Agricultural Biotechnology: Country Case Studies – a Decade of Development
*Edited by G.J. Persley and L.R. MacIntyre*

* Out of print

*Biotechnology in Agriculture No. 25*

# Agricultural Biotechnology: Country Case Studies –

## a Decade of Development

---

*Edited by*

**Gabrielle J. Persley,**
*Chair, The Doyle Foundation, Glasgow, Scotland*

**L. Reginald MacIntyre**
*Formerly Director, Communications Division, International Development Research Centre, Ottawa, Canada*

CABI *Publishing*

**CABI *Publishing* is a division of CAB *International***

CABI Publishing
CAB International
Wallingford
Oxon OX10 8DE
UK

Tel: +44 (0) 1491 832111
Fax: +44 (0) 1491 833508
Email: cabi@cabi.org
Web site: www.cabi-publishing.org

CABI Publishing
10 E 40th Street
Suite 3203
New York, NY 10016
USA

Tel: +1 212 481 7018
Fax: +1 212 686 7993
Email: cabi-nao@cabi.org



A catalogue record for this book is available from the British Library, London, UK.

A catalogue record for this book is available from the Library of Congress, Washington, DC, USA.

ISBN 0 85198 816 4

Typeset by Wyvern 21 Ltd, Bristol
Printed and bound in the UK by Cromwell Press, Trowbridge

# Contents

# Contributors

**Chetsanga, C.J.** Biotechnology Research Institute, 1574 Alpes Road, Hatcliffe, PO Box 6640, Harare, Zimbabwe

**Dart, P.J.** Department of Agriculture, University of Queensland, St. Lucia, Queensland 4000, Australia

**de la Cruz, Reynaldo E.** National Institutes of Molecular Biology and Biotechnology (BIOTECH), University of the Philippines, Los Baños, Philippines

**Espinoza, Ana Mercedes** Centro de Investigacion en Biologia Celular y Molecular, Universidad de Costa Rica, San José, Costa Rica

**Falconi, Cesar** Inter-American Development Bank, 1300 New York Avenue NW, Washington, DC 20577, USA

**Madkour, Magdy A.** Agricultural Genetic Engineering Research Institute, Gaioa Street, Giza 12619, Cairo, Egypt

**Munoz, Miguel** Centro de Investigacion en Biologia Celular y Molecular, Universidad de Costa Rica, San José, Costa Rica

**Olembo, N.K.** Industrial Property Office, Ministry of Education, Science and Technology, Nairobi, Kenya

**Persley, Gabrielle J.** The Doyle Foundation, 45 St Germains, Bearsden, Glasgow, G61 2RS, Scotland, UK

**Sampaio, Maria Jose Amstalden** Embrapa-Labex; and US Plant, Soil and Nutrition Lab-ARS/USDA, Tower Road, Ithaca, NY 14853–2901, USA

**Sharma, Manju** Department of Biotechnology, Ministry of Science and Technology, Block 2 (7th floor), CGO Complex, Lodi Road, New Delhi 110 003, India

**Sittenfeld, Ana** Centro de Investigacion en Biologia Celular y Molecular, Universidad de Costa Rica, San José, Costa Rica

**Slamet-Loedin, I.H.** R & D Center for Biotechnology, Indonesian Institute of Sciences, Jl. Raya Bogor Km. 46, Cibinong 16911, Indonesia

**Sukara, E.** R & D Center for Biotechnology, Indonesian Institute of Sciences, Jl. Raya Bogor Km. 46, Cibinong 16911, Indonesia

**Tanticharoen, Morakot** National Center for Genetic Engineering and Biotechnology, 73/1 Rama 6 Road, Ratchadevi, Bangkok 10400, Thailand

**Torres, Ricardo** Carrera 20 No. 86A-09 (302), Bogota, Colombia

**Zafar, Yusuf** Plant Biotechnology Division, National Institute for Biotechnology and Genetic Engineering, Faisalabad, Pakistan

**Zamora, Alejandro** Centro de Investigacion en Biologia Celular y Molecular, Universidad de Costa Rica, San José, Costa Rica

**Zhang, Qifa** Huazhong Agricultural University, National Key Laboratory of Crop Genetic Improvement, Wuhan 430070, China

# Foreword

The 1990–2000 decade has seen extraordinary advances in all areas of biotechnology, and international agriculture has benefited from this new and exciting 'promethean' science. This term conveys the 'daringly original and creative' science that is necessary to deal with the problems of poverty, hunger and environmental degradation facing much of the world today.

A 1990 CAB *International* publication (*Beyond Mendel's Garden: Biotechnology in the Service of World Agriculture*) and a 2000 publication on *Promethean Science: Agricultural Biotechnology, the Environment and the Poor* summarize the scientific and policy developments in agricultural biotechnology over the past decade. These and other publications by CAB *International* and the Consultative Group on International Agricultural Research looked at the future role of biotechnology in agriculture. They helped set the agenda for a decade of exciting research, particularly in developing countries and by the international agricultural research centres. They also foreshadowed the biotechnology debate that continues today on biosafety issues, effects on biodiversity and the environment and intellectual property rights. Despite the ongoing controversies about genetically improved foods, we can look back on the 1990s as a decade of great scientific achievements and practical applications of science.

This publication, through its overview chapter and selected country studies, contains useful information on the evolving biotechnological research in the main geographical regions. The emphasis is on the potential these new technologies hold for agriculture in developing countries. The reports vary in the depth of coverage, but all combine to show the urgent need that exists for public- and private-sector investment to ensure that all countries share in the benefits of modern biotechnologies, while minimizing any unintended effects.

Ismail Serageldin
Library of Alexandria

# Preface

Biotechnology is having a profound impact on agricultural research and productivity, and developing countries need to organize their national research systems accordingly. Expertise in the basic biological sciences is still in short supply in many developing countries, though great strides have been made in training during the past decade. Mobilization of the basic science laboratories usually found in the universities to solve agricultural problems will require new policy and institutional arrangements and additional resources. Most countries also need to upgrade their teaching of biological sciences at all levels.

The increasing dominance of private-sector research and development (R & D) in agricultural biotechnology may increase the cost of access by the developing countries to advances in science and technology, which were previously freely available as public goods. The international agricultural research centres (IARCs) may assume greater importance as points of access to advanced technologies for the national agricultural research systems (NARS), especially for the many smaller countries (about 70).

Developing countries are at different stages in determining their policies and programmes in relation to biotechnology. They fall into three main categories: (i) countries with interest but no direct involvement in biotechnology; (ii) countries with a national policy and programme, most of which are concerned with traditional biotechnology and with collaborative links overseas; and (iii) countries with a national policy and programme, including an active in-country programme on modern biotechnology. In this volume we have selected country case studies that reflect these three categories.

The status of biotechnology has been assessed over several years in the countries included here. Some of the countries included have well-formulated policies, national programmes and adequate financial resources. Others, although committed to biotechnology at a policy level, have not formulated a national programme and do not yet have significant financial support for their activities. The development of a well-defined national biotechnology strategy and a plan of activities that clearly delineates priorities for the public and private sector is a critical step for the implementation of an effective national programme in biotechnology that makes efficient use of domestic and external resources.

International transfer of biotechnology is under way but could be considerably strengthened by international agencies and bilateral agreements. To date, it has been largely limited to the transfer of technology from the public sector in industrial countries to the public sector in developing countries. Given the substantial private-sector investments in biotechnology in industrial countries, it is likely that the private sector has new technologies that could usefully be applied or adapted to agriculture in developing countries. The challenge to the international development agencies, the developing countries and the IARCs is to devise innovative mechanisms to facilitate the transfer of technology from the private sector in industrial countries to the public and private sector in the developing countries, under mutually beneficial arrangements.

Important issues now on the international agenda include biodiversity, biosafety and environmental concerns. Many countries have taken up these issues, and some have developed standards for the safe handling and movement of genetically altered plants and animals. All countries must be concerned about the conservation of their natural resources.

National governments and international development agencies, such as the World Bank, have published guidelines to address these areas and to inform the community about the enormous potential offered by modern science to produce the food, fuel and fibre needed to double food production by the year 2020 to meet population growth.

This publication brings to a wide international audience useful information on the status of biotechnological research in a selected group of countries. It will hopefully stimulate wider public debate, which will add to the growing body of knowledge in this sometimes controversial area of modern agricultural research.

Gabrielle J. Persley
L. Reginald MacIntyre

# Acknowledgements

Many people contributed to this publication and we are grateful for their valuable input. The authors of the various chapters are specially thanked for their thorough research and detailed presentations.

The editors are grateful for the financial support of those sponsoring institutions – the International Service for National Agricultural Research, the World Bank, the Australian International Development Agency (AusAID), and the Australian Centre for International Agricultural Research – which supported the commissioning of some of the initial country studies that have been further developed here. We are grateful to the Consultative Group on International Agricultural Research (CGIAR) for allowing the generous use of up-to-date material published in *Agricultural Biotechnology and the Poor*, published by the CGIAR in January 2000. We also thank the authors of that material for agreeing to be contributors to this book.

The late Dr John J. Doyle provided inspiration and input, particularly with regard to livestock biotechnology. Pamela George, formerly of the World Bank in Washington, provided valuable assistance throughout the preparation of this monograph. We are also grateful to Ida MacIntyre, who provided technical editing and word-processing skills, which together speeded up the completion of this publication.

Ms Carlene Brenner did a thorough technical review of the papers, and advised on selection of material for inclusion in the book. We thank also Professor Declan McKeever for his advice on livestock diseases in Kenya.

And, finally, we would like to thank Tim Hardwick of CABI *Publishing* for his patience and guidance throughout the writing and revision phases of this book.

# Acronyms and Abbreviations

| | |
|---|---|
| AARD | Agency for Agricultural Research and Development (Indonesia) |
| ABP | Agricultural Biotechnology Programme (Colombia) |
| ACIAR | Australian Centre for International Agricultural Research |
| ADB | Asian Development Bank |
| ADC | African Development Corporation |
| AFMA | Agriculture Fisheries Modernization Set (Philippines) |
| AGERI | Agricultural Genetic Engineering Research Institute (Egypt) |
| ARIPO | African Regional Intellectual Property Organization (Zambia) |
| AusAID | Australian International Development Assistance Bureau |
| AUSTRADE | Australian Department of Overseas Trade |
| AVRDC | Asian Vegetable Research and Development Centre (Taiwan) |
| BIOTEC | National Centre for Genetic Engineering and Biotechnology (Thailand) |
| BIOTROP | South-East Asian Regional Centre for Tropical Biology (Bogor, Indonesia) |
| BPPT | Agency for Technology Assessment and Application (Indonesia) |
| BRI | Biotechnology Research Institute (Zimbabwe) |
| Bt | *Bacillus thuringiensis* |
| CBI | Crop Breeding Institute (Zimbabwe) |
| CEMB | Centre of Excellence in Molecular Biology (Pakistan) |
| CENICAFE | National Coffee Research Centre (Colombia) |
| CENICAÑA | Colombian Sugarcane Research Centre |
| CEPAL | Comision Economica para America Latina y el Caribe |
| CFTC | Commonwealth Fund for Technical Cooperation (UK) |
| CGIAR | Consultative Group on International Agricultural Research |

| | |
|---|---|
| CIAT | Centro Internacional de Agricultura Tropical (International Centre for Tropical Agriculture) (Colombia) |
| CIBCM | Center for Research in Cellular and Molecular Biology |
| CIFOR | Centre for International Forestry Research (Indonesia) |
| CIMMYT | Centro Internacional de Mejoramiento de Maiz y Trigo (International Centre for the Improvement of Wheat and Maize) (Mexico) |
| CIP | Centro Internacional de la Papa (International Potato Centre) (Peru) |
| CIRAD | Centre de Cooperation Internationale en Recherche Agronomique pour le Développement |
| CMV | cucumber mosaic virus |
| COLCIENCIAS | Colombian Institute for Science and Technology |
| COLTABACO | Colombian Tobacco Company |
| CONPES | National Council for Social Economic Policy (Colombia) |
| CORPOICA | Agricultural Research Colombian Corporation |
| COST | The ASEAN Committee on Science and Technology |
| CRIF | Central Research Institute for Fisheries (Indonesia) |
| CRIFC | Central Research Institute for Food Crops (Indonesia) |
| CRIH | Central Research Institute for Horticulture (Indonesia) |
| CRIIC | Central Research Institute for Industrial Crops (Indonesia) |
| CSC | Commonwealth Science Council (UK) |
| CSIRO | Commonwealth Scientific and Industrial Research Organization (Australia) |
| DFID | Department for International Development (UK) |
| DRSS | Department of Research and Specialist Services (Zimbabwe) |
| ECB | Colombian Biotechnology Company |
| ECF | East Coast fever |
| EEC | European Economic Commission |
| ELISA | enzyme-linked immunosorbent assay |
| FAO | Food and Agriculture Organization of the United Nations |
| FINNIDA | Finnish International Development Agency |
| GIO | genetically improved organism |
| GM | genetically modified |
| GMO | genetically modified organism |
| GSL | Green Seed Ltda. (Colombia) |
| GTZ | Deutsche Gesselschaft für Technische Zusammenarbeit (Germany) |
| IARC | international agricultural research centre |
| IBUN | Institute of Biotechnology, Universidad Nacional de Colombia |
| ICA | Instituto Colombiano Agropecuaria (Columbian Agricultural Institute) |
| ICIPE | International Centre of Insect Physiology and Ecology (Kenya) |

| | |
|---|---|
| ICRAF | International Centre for Research in Agroforestry (Kenya) |
| ICRISAT | International Crops Research Institute for the Semi-Arid Tropics (India) |
| IDRC | International Development Research Centre (Canada) |
| ILRI | International Livestock Research Institute (Kenya) |
| INBio | Instituto Nacional de Biodiversidad (Costa Rica) |
| IPB | Bogor Agricultural University (Indonesia) |
| IPGRI | International Plant Genetic Resources Institute (Rome) |
| IPM | Integrated Pest Management |
| IPR | intellectual property rights |
| IRRI | International Rice Research Institute (Philippines) |
| ISNAR | International Service for National Agricultural Research (the Netherlands) |
| ITB | Bandung Institute of Technology (Indonesia) |
| IUC | Inter-University Centres (Indonesia) |
| JICA | Japan International Cooperation Agency |
| KARI | Kenya Agricultural Research Institute |
| KEFRI | Kenya Forestry Research Institute |
| KEMFRI | Kenya Marine Fisheries Research Institute |
| KETRI | Kenya Trypanosomosis Research Institute |
| LEHRI | Lembang Research Institute for Horticulture (Indonesia) |
| LIPI | Indonesian Institute of Sciences |
| LMO | living modified organism |
| MAS | marker-assisted selection |
| MINAE | Ministry of the Environment and Energy (Costa Rica) |
| NARAD | Norwegian Agency for International Development |
| NARS | National agricultural research systems |
| NCGEB | National Centre for Genetic Engineering and Biotechnology (Thailand) |
| NCL | National Chemistry Laboratory (India) |
| NCST | National Council for Science and Technology (Kenya) |
| NIAB | Nuclear Institute for Agriculture and Biology (Pakistan) |
| NECTEC | National Electronics and Computer Technology Centre (Thailand) |
| NIBGE | National Institute for Biotechnology and Genetic Engineering (Pakistan) |
| NSTDA | National Science and Technology Development Agency (Thailand) |
| OECD | Organization for Economic Cooperation and Development |
| ORSTOM | Office de Recherche Scientifique et Technique Outre Mer (France) |
| PPPKT | Research and Development Centre for Applied Chemistry (Bandung, Indonesia) |
| PSTV | peanut stripe virus |
| PUJ | Plant Biotechnology Unit, Pontificia Universidad Javeriana |

| | |
|---|---|
| QTL | quantitative trait locus/loci |
| RAPD | ramdomly amplified polymorphic DNA |
| RCZ | Research Council of Zimbabwe |
| RDCBt | Research and Development Centre for Biotechnology (Indonesia) |
| RFLP | restriction fragment length polymorphism |
| RHBV | rice hoja blanca virus |
| RIA | radioimmunoassay |
| RIAD | Research Institute for Animal Diseases (Indonesia) |
| RIAP | Research Institute for Animal Production (Indonesia) |
| SIRDC | Scientific and Industrial Research and Development Centre (Zimbabwe) |
| TDRI | Thai Development Research Institute |
| TMV | tomato mosaic virus |
| TRIPs | Trade-Related Aspects of Intellectual Property Rights |
| UGM | University of Gadjah Mada (Indonesia) |
| UNDP | United Nations Development Programme |
| UNEP | United Nations Environment Programme |
| UNIDO | United Nations Industrial Development Organization |
| UPLB | University of the Philippines Los Baños |
| UPOV | Union Internationale pour la Protection des Obtentions Végétales (Convention for the Protection of New Varieties of Plants) |
| USAID | United States Agency for International Development |
| WHO | World Health Organization |
| WIPO | World Intellectual Property Organization |
| WTO | World Trade Organization |

# Dedication

**This book is dedicated to**

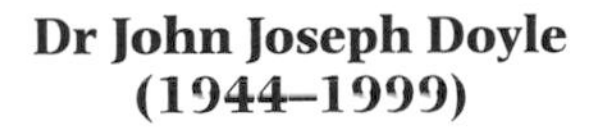

**Dr John Joseph Doyle**
**(1944–1999)**

Formerly Deputy Director General,
International Laboratory for Research on Animal Diseases, Nairobi, Kenya.

Jack Doyle strongly believed that science could solve human problems. He did pioneering work in molecular biology in Africa from 1975 to 1995, and was one of the main architects responsible for the establishment of ILRAD (now part of the International Livestock Research Institute) as a first-class institute for molecular biology in Africa. He was especially concerned with the control of livestock diseases, including trypanosomosis and East Coast Fever. He was a strong supporter of the crucial role of scientists in national research systems and universities, and worked with scientists in many of the countries reviewed in this volume. His ideals are being pursued by his colleagues and friends through the work of the Doyle Foundation established in June 2000 (www.doylefoundation.org). Its purpose is to support the role of science in international development, through fellowships and other activities.

# I

# Introductory Review

# 1

# Agricultural Biotechnology: Global Challenges and Emerging Science

**Gabrielle J. Persley**

In global terms, increases in world food production have more than kept pace with the increases in the global population to date. Although the world agricultural growth rate has decreased from 3% in the 1960s to 2% in the past decade, the aggregated projections show that, given reasonable initial assumptions, world food supply will continue to outpace world population growth, at least to 2020. Worldwide, per capita availability of food is projected to increase around 7% between 1993 and 2020 (IFPRI, 1997). Therein lies a paradox.

The first aspect of the paradox is that despite the increasing availability of food, approximately 800 million people out of the global population of 6 billion are food insecure. They dwell among the 4.5 billion inhabitants of Asia (48%), Africa (35%) and Latin America (17%). Of these 800 million people, a quarter are malnourished children (IFPRI, 1997).

Children and women are most vulnerable to dietary deficiencies. Dietary micronutritional deficiencies accompany malnutrition. Vitamin A deficiency is prevalent in the developing countries and it is estimated that over 14 million children under 5 years of age suffer eye damage as a result. Up to 4% of severely affected children will die within months of going blind and even mild deficiencies can significantly increase mortality rates in children. Iron deficiency affects 1 billion people in the developing world, particularly women and children and its effects are compounded by common tropical diseases.

The second aspect of the paradox is that food insecurity is so prevalent at a time when global food prices are generally in decline. Over the period 1960–1990, world cereal production doubled, per capita food production increased 37%, calories supplied increased 35% and real food prices fell by almost 50% (McCalla, 1998).

The basic cause of the paradox is the intrinsic linkage between poverty and food security. Simply put, people's access to food depends on income.

Poverty is both a rural and an urban phenomenon. Over 1.3 billion people in developing countries are absolutely poor, with incomes of US$1 per day or less per person, while another 2 billion people live on less than US$2 per day (World Bank, 1997). Most live in the low-potential, rain-fed rural areas of the world. With increasing urbanization, a higher proportion will be living in the cities of the developing countries by the mid-21st century. Ensuring their access to sufficient nutritious food at affordable prices is also an important component of global food-security strategies. Agricultural research needs to respond to both of these challenges, so as to improve the livelihood of families who live in rural areas and ensure the increased availability of nutritious food at affordable prices for the urban dwellers.

## Global Challenges

The most important global challenges are as follows:

- Reducing poverty, especially in rural areas.
- Improving food security and reducing malnutrition.
- Providing sufficient income for the rapidly increasing numbers of urban poor.
- Mobilizing new technologies for environmentally sustainable development.

The global problems facing agriculture are described by Swaminathan (2000):

- First, increasing population leading to increased demand for food and reduced per capita availability of arable land and irrigation water.
- Secondly, improved purchasing power and increased urbanization leading to higher per capita food grain requirements due to an increased consumption of animal products.
- Thirdly, marine fish production is becoming stagnant.
- Fourthly, there is increasing damage to the ecological foundations of agriculture, such as land, water, forests and biodiversity, as well as climate change.
- Finally, while dramatic new technological developments are taking place, particularly in biotechnology, their environmental and social implications are yet to be fully understood.

### World food production challenge

*Production trends*

As a result of the green revolution, yields of maize, wheat and rice in developing countries doubled between 1961 and 1991. In Africa, the annual increase in yield per hectare for maize, wheat and rice (1.3%) is less than a third of that achieved in Asia (4.5%). This presents a significant opportunity to raise cereal production in Africa through yield increases. Food-animal production in developing countries increased by 15% in the 1980s, whereas the global increase was 24%.

*Consumption patterns*

Demand for food in developing countries is met by both local production and imports. Currently developing countries are net importers of 88 million tons of cereals per year at a cost of US$14.5 billion. In Africa, the current annual production shortfalls of cereals, met by imports, is about 21.5 million tons. It is predicted that these shortfalls will increase tenfold by 2025. Since the 1970s, the developing countries have been increasingly large net importers of milk and meat (except pig meat).

*Future food demands*

There will be a global demand for 40% more grain in 2020, with most of the demand coming from developing countries. This will include a doubling in demand for feed grains in developing countries. Net cereal imports by developing countries will almost double to meet the gap between production and demand (IFPRI, 1997).

## Future challenges

The food production increases over the past 40 years have been achieved largely by increasing productivity of cereals, expanding the area of arable land and massive increases in fertilizer and pesticide use. To meet the production challenges of the next decades, there is a need to:

- Increase biological yields of the major food crops.
- Improve productivity of livestock.
- Improve nutrient content in the diet, especially of women and children.
- Intensify agriculture, since land for agriculture is increasingly scarce.
- Manage natural resources in a sustainable way.

*Environmental trends*

- The intensification of agriculture in the favourable areas has come at the cost of damage to the environment, with increasing salinity problems in irrigated areas and damage to human health and wildlife due to misuse of pesticides.
- Decreasing water availability for agriculture is one of the most important trends over the last decade. There is a need for more efficient use of water in agriculture, including the development of water-saving and drought-tolerant genotypes and more efficient water-management practices.
- Pressure on land from urbanization and industrialization increases. There are limited prospects for expanding the land available for agriculture, except by moving into forests or marginal areas. The latter have poor soils and little irrigation and are prone to drought.

- Deforestation and loss of biodiversity are caused by the clearing of land for logging in areas of terrestrial mega-biodiversity. The use of modern plant varieties also threatens the loss of landraces of crops.
- Natural disasters pose a continuing threat and the long-term effects of climate change are unknown.

*Trade and competitiveness trends*

- Increasing trade: one option for obtaining the necessary purchasing power for food is through increased interregional and international trade.
- Increasing competitiveness: the declining prices for agricultural commodities suggest a need also to increase the productivity of agricultural exports and to develop new value-added products for export.
- Product quality needs to meet the certification and food safety standards of importing countries.

## Food security strategies

While there is a need for further intensification of agricultural production to meet projected demand for food, intensification strategies must change in order to avoid an adverse environmental impact and reverse the effects of past practices (Pinstrup-Andersen and Cohen, 2000). Strategies to achieve the needed increases in food supply over the next 25 years include the following:

- Sustainable productivity increases in food, feed and fibre crops in both irrigated and rain-fed areas.
- Reducing chemical inputs of fertilizers and pesticides and replacing these with biologically based products.
- Integrating soil, water and nutrient management.
- Improving the nutrition and productivity of livestock and controlling livestock diseases.
- Sustainable increases in fisheries and aquaculture production.
- Increasing trade and competitiveness in global markets.

The challenge now is how to use new developments in modern science, including biotechnology, together with new information and communications technology and new ways of managing knowledge to make complex agricultural systems more productive in sustainable ways.

As Swaminathan (2000) says,

> we need to examine how science can be mobilized to raise further the biological productivity ceiling without associated ecological harm. Scientific progress on the farms, as an ever-green revolution, must emphasize that the productivity advance is sustainable over time since it is rooted in the principles of ecology, economics, social and gender equity and employment generation.

## Promethean Science

The pace of change in modern science led Persley (2000) to term it 'Promethean [daringly original and creative] science', acknowledging both its risks and benefits. The term stems from the Greek Titan, Prometheus, who legend says introduced fire to humans. Biotechnology, like fire, carries with it benefits and risks, depending on its use (see also Serageldin and Persley, 2000).

Modern science encompasses new developments in the biological, physical and social sciences. In the biological sciences, discoveries over the past 20 years allow much greater understanding of the structure and function of human, animal and plant genes. At the same time, new discoveries in the physical sciences underpin the revolution in information and communications technologies. For example, the use of geographical information systems enables characterization of agroecosystems and also offers means by which new technologies can be customized to the needs of particular agroecosystems. The biological and physical sciences also interact in new ways. For example, the ability to analyse large volumes of data is a critical component of various genome projects that are mapping all the genes in an organism, as in the Human Genome Project (Doyle and Persley, 1998).

There have also been new developments in the social sciences that underpin community participation in technology development and evaluation (sometimes termed agroecological methods). Participatory methods developed in the social sciences can help in the understanding of the problems and the researchable issues, particularly of small farmers operating in marginal environments. They may also be used to clarify the concerns of people in rural and urban communities in regard to the deployment of new technologies, including the products of biotechnology.

It is the successful integration of all branches of modern science and traditional knowledge that is required to develop knowledge-intensive solutions to the problems of rural communities. These solutions need to be not only technically feasible but also socially acceptable. Indeed, the potential value of modern science to agriculture and the environment will not be realized without major additional efforts involving all stakeholders, including civil society, farmer cooperatives, producers, consumers, governments and development agencies.

### Scope of biotechnology

Biotechnology broadly defined refers to any technique that uses living organisms or substances from these organisms to make or modify a product, improve plants or animals or develop microorganisms for specific uses. Biotechnology consists of a gradient of technologies, ranging from the long-established and widely used techniques of traditional biotechnology (for example, food fermentation and biological control), through to novel and continuously evolving techniques of modern biotechnology (Fig. 1.1).

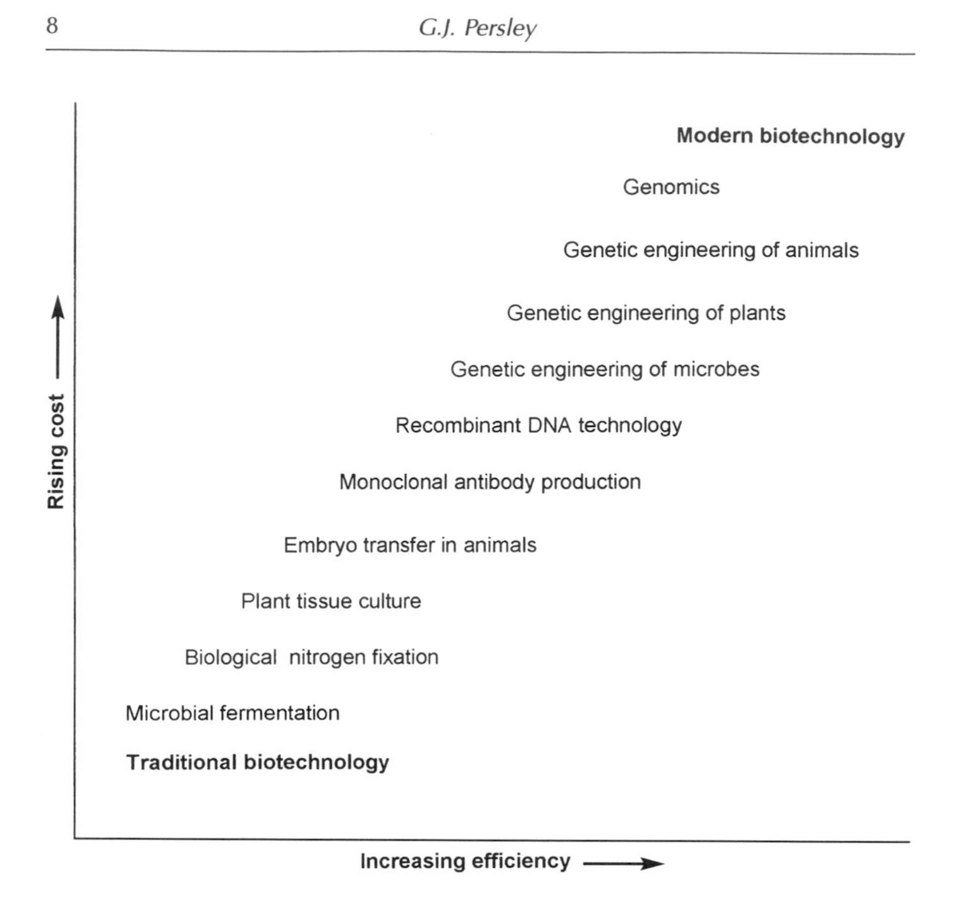


**Fig. 1.1.** Gradient of biotechnologies (from Persley, 1990; Doyle and Persley, 1996).

During the 1970s scientists developed new methods for precise recombination of portions of deoxyribonucleic acid (DNA), the biochemical material in all living cells that conveys the instructions that govern inherited characteristics, and for transferring portions of DNA from one organism to another. This set of enabling techniques is referred to as recombinant DNA technology or genetic engineering. Modern biotechnology currently includes the various uses of new techniques of recombinant DNA technology, monoclonal antibodies and new cell and tissue culture methods.

Over the past two decades the number of significant advances made in modern biotechnology has increased dramatically. It is this increase in the use of new techniques for understanding and modifying the genetics of living organisms that has led to greatly increased interest and investment in biotechnology. This has been accompanied by increasing concerns as to the power of the new technologies and the safety of their use, both to human health and to the environment.

## Evolution of Modern Genetics

Mendel's laws of genetics were rediscovered in 1900. Mendel had published his original work on inheritance patterns in pea in 1865, but it took 35 years for others to grasp their significance. Since 1900, we have witnessed steady progress in our understanding of the genetic make-up of all living organisms, ranging from microbes to humans. A major step in human control over genetic traits was taken in the 1920s when Muller and Stadler discovered that radiation can induce mutations in animals and plants.

In the 1930s and 1940s, several new methods of chromosome and gene manipulation were discovered and used in plant improvement. These included the use of colchicine to achieve a doubling in chromosome number, commercial exploitation of hybrid vigour in maize and other crops, use of chemicals such as nitrogen mustard and ethyl methane sulphonate to induce mutations and techniques such as tissue culture and embryo rescue to get viable hybrids from distantly related species.

The double-helix structure of DNA, the chemical substance of heredity, was discovered in 1953 by James Watson and Francis Crick. This, combined with the development of recombinant DNA technology, triggered explosive progress in every field of genetics. Today, there is a rapid transition from Mendelian to molecular genetic applications in agriculture, medicine and industry.

This brief review of genetic progress from 1900 to 1999 (see Swaminathan, 2000) stresses that knowledge and discovery represent a continuum, with each generation taking our understanding of the complex web of life to a higher level. It would therefore be unwise to either adopt or discard experimental tools or scientific innovations because they are either old or new. Just as it took 35 years for biologists to understand fully the significance of Mendel's work, it may take a decade or more to understand fully the benefits and risks associated with living modified organisms (LMOs), including genetically improved foods (Persley and Siedow, 1999).

### The gene revolution

The 1990s have seen dramatic advances in our understanding of how biological organisms function at the molecular level, as well as in our ability to analyse, understand and manipulate DNA molecules, the biological material from which the genes in all organisms are made. The entire process has been accelerated by the Human Genome Project, which has invested substantial public and private resources into the development of new technologies to work with human genes. The same technologies are directly applicable to all other organisms, including plants and animals. The first complete sequence of a plant genome – the flowering plant *Arabidopsis thaliana* – is now completed (Arabidopsis Genome Initiative, 2000). The rice genome is also very close to completion. The new scientific discipline of genomics has arisen, which has contributed to powerful new approaches to identify the structure and functions of genes and their applications in human health, agriculture and the environment. These new discoveries and their commercial applications have helped to promote the biotechnology industry, mainly in North America and Europe.

## Commercial Applications of Agricultural Biotechnology

The greater specificity in the handling of genes since the 1970s has meant that inventors could protect their discoveries by means of patents and other forms of intellectual property rights (IPR). This has led to an explosion of private investment in the biosciences, leading to what has been called a biotechnology revolution. Most modern biotechnology applications are in health care, where they offer new hope to patients with AIDS, genetically inherited diseases, diabetes, influenza and some forms of cancer. Biotechnology-based processes are now used routinely in the production of most new medicines, diagnostic tools and medical therapies. The global market for these products is approximately US$13 billion.

New developments in agricultural biotechnology are being used to increase the productivity of crops, primarily by reducing the costs of production by decreasing the needs for inputs of pesticides and herbicides, mostly in crops grown in temperate zones. The applications of agricultural biotechnology are developing new strains of plants that give higher yields with fewer inputs, can be grown in a wider range of environments, give better rotations to conserve natural resources and provide more nutritious harvested products that keep much longer in storage and transport, and continued low-cost food supplies for consumers.

Private industry has dominated research, accounting for approximately 80% of all R & D. Consolidation of the industry has proceeded rapidly since 1996, with more than 25 major acquisitions and alliances, worth US$15 billion.

During the past decade, the commercial cultivation of transgenic plant varieties became well established, particularly during the latter part of the decade. In 1999, it is estimated that approximately 40 million ha of land were planted with transgenic varieties of over 20 plant species, the most commercially important of which were cotton, maize, soybean and rape-seed (James, 1999). The value of the global market in transgenic crops grew from US$75 million in 1995 to US$1.64 billion in 1998.

The traits these new varieties contain include insect resistance (cotton, maize), herbicide resistance (maize, soybean) and delayed fruit ripening (tomato). The benefits of these new crops are better weed and insect control, higher productivity and more flexible crop management. These benefits accrue primarily to farmers and agribusinesses, although there are also economic benefits accruing to consumers in terms of maintaining food production at low prices. Health benefits for consumers are also emerging from new varieties of maize and rape-seed with modified oil content and reduced levels of potentially carcinogenic mycotoxins. The broader benefits to the environment and the community through reduced use of pesticides contribute to more sustainable agriculture and improved food security.

Other crop/input trait combinations currently being field-tested include virus-resistant melon, papaya, potato, squash, tomato and sweet pepper; insect-resistant rice, soybean and tomato; disease-resistant potato; and delayed-ripening chilli pepper. There is also work in progress to use plants such as maize, potato and banana as mini-factories for the production of vaccines and biodegradable plastics.

Several large corporations in Europe and the USA have made major investments to adapt the new discoveries in the biological sciences to commercial purposes, especially to produce new plant varieties of agricultural importance for large-scale commercial agriculture. The same technologies also have important potential applications to address problems of poverty reduction, food security, environmental conservation and trade competitiveness in developing countries (Tzotzos and Skryabin, 2000; US National Academy of Sciences, 2000).

## Scientific advances

Further scientific advances will probably result in crops with a wider range of traits, some of which are likely to be of more direct interest to consumers – for example, by having traits that confer improved nutritional quality in food. Crops with improved output traits could have nutritional benefits for millions of people who suffer from malnutrition and deficiency disorders. Genes have been identified that can modify and enhance the composition of oils, proteins, carbohydrates and starch in food/feedgrains and root crops. A gene encoding β-carotene/vitamin A formation, for example, has been incorporated experimentally in rice. This is being evaluated for the feasibility of using vitamin-A enriched rice to enhance the diets of the 180 million children who suffer from vitamin A deficiency. Similarly, introducing genes that increase available iron levels in rice threefold is a potential remedy for iron deficiency, which affects more than 2 billion people and causes anaemia in about half that number.

Applications of biotechnology in agriculture are in their infancy. Most current genetically engineered plant varieties are modified only for a single trait, such as herbicide tolerance or pest resistance. The rapid progress being made in genomics may enhance plant breeding as the functions of more genes and how they control particular traits are identified. This may enable more successful breeding for complex traits, such as drought and salt tolerance. This would be of great benefit to those farming in marginal and rain-fed lands worldwide, because breeding for such traits, has had limited success with conventional breeding of the major staple food crops.

## Applications of Biotechnology for Poverty Reduction and Food Security

Breakthroughs in modern science have led to rapid progress in understanding the genetic basis of living organisms and the ability to use that understanding to develop new products and processes useful in human and animal health, food and agriculture and the environment. The adoption of modern biotechnology is most advanced in human health, where many new drugs, diagnostics and vaccines are based on the use of new biotechnology.

The applications developed from the new methods in biotechnology place them within the continuum of techniques used throughout human history in industry, agriculture and food processing. Thus, while modern biotechnology provides powerful new tools, these tools are used to generate products that in most cases fill similar roles to those produced with more traditional methods.

In agriculture, there is now increasing use of modern molecular genetics for genetic mapping and marker-assisted selection as aids to give more precise and rapid development of new strains of improved crops, livestock, fish and trees. Other biotechnology applications, such as tissue culture and micropropagation, are being used for the rapid multiplication of horticultural crops and trees. New diagnostics and vaccines are being widely adopted for the diagnosis, prevention and control of fish and livestock diseases.

Harnessing the full power of the genetic revolution requires going beyond these early applications of modern biotechnology and recognizing the power of the new revolution in genomics and associated technologies as aids for genetic improvement. The new technologies enable greatly increased efficiency of selection for useful genes, based on knowledge of the biology of the organism, the function of specific genes and their role in regulating particular traits. This will enable more precise selection of improved strains. These techniques may be used for more efficient selection in conventional breeding programmes. They may also be used for the identification of genes suitable for use in the development of transgenic strains (Serageldin and Persley, 2000).

The most striking differences between the techniques of modern biotechnology and those that have been used for many years lie in the increased precision with which the new techniques may be used and the shorter time required to produce results. For example, modern biotechnology enables plant breeders to collaborate with molecular biologists and transfer to a popular and highly developed crop variety one or two specific genes needed to impart a new characteristic, such as a specific kind of pest resistance.

The advantages of the new techniques of modern biotechnology are that they speed up plant and animal breeding, offer possible solutions to previously intractable problems and difficult targets, such as drought tolerance, and

**Table 1.1.** Illustrative applications of biotechnology to the goal of poverty reduction.

| | | | Biotechnology applications (1–6) | | | | | | |
|---|---|---|---|---|---|---|---|---|---|
| | | | 1 | 2 | 3 | 4 | 5 | 6 | 7 |
| Poverty reduction objective | Constraint | Target | Microbial fer-mentation/ biocontrol/ biofertilizers | New diag-nostics/ vaccines | Tissue culture/ microprop-agation | Molecular markers and MAS | Genetic engineering/ transgenics | Genomics | Examples |
| Increasing rural incomes | Lack of clean seed/ planting material | Vegetative crops and trees | Bio-pesticides | Plant disease diagnostics | Cardamom | | | | India |
| | | | | | Potato | | | | Vietnam |
| | | | | | Banana | | | | Kenya |
| Sustainable production in resource-poor areas | Drought | Cereals | | | | Maize | | Drought tolerance in cereals | AMBN/ CIMMYT |
| | Pests | Maize | | | | Insect resistance | | | |
| | Acid soils | Aluminium tolerance | | | | Aluminium tolerance | | | |
| More nutritious food | Vitamins | Rice | | | | | Vit. A rice | | IRRI |
| | Micro-nutrients | Rice | | | | Iron | | | India China |

Notes:
1. Traditional biotechnology applications, such as microbial and food fermentation.
2. New diagnostics and vaccines based on molecular applications.
3. New methods for tissue culture and micropropagation of planting material.
4. Use of molecular markers in marker-assisted selection (MAS) in conventional plant and animal breeding.
5. Genetic engineering to produce transgenic plant and/or animal strains, containing new specific gene(s) controlling a particular trait.
6. Genomics: understanding the physical structure of the genome and, in functional genomics, the function of specific genes.
7. Specific global examples where new biotechnologies are being developed.

AMBN, Asian Maize Biotechnology Network; CIMMYT, Centro Internacional de Mejoramiento de Maiz y Trigo; IRRI, International Rice Research Institute.

**Table 1.2.** Illustrative applications of biotechnology to the goal of food security.

| Food security objective | Constraint | Target | Biotechnology applications (1–6) | | | | | | |
|---|---|---|---|---|---|---|---|---|---|
| | | | 1 Microbial fermentation/ biocontrol/bio-fertilizers | 2 New diag-nostics/ vaccines | 3 Tissue culture/ microprop-agation | 4 Molecular markers and MAS | 5 Genetic engineering/ transgenics | 6 Genomics | 7 Examples |
| Meeting demand predictions for staple foods | Pests/ diseases | Rice | Rice FAO/IPM programme | | | | Bacterial blight (Xa1) | Rice genome | ARBN/ IRRI China |
| | Abiotic stresses | Salinity/ drought tolerance in cereals | | | | Cereals/ drought tolerance | Salinity tolerance (mangrove gene) | Cereals | India China ICRISAT CIMMYT |
| Increasing livestock | Diseases | Cattle/pigs/ sheep | | FMD | | | | | Thailand |
| | Productivity | Dairy cattle | | | Embryo technology | | | | India |
| Increasing fish/aqua-culture | Diseases | Shrimp viruses | | Molecular diagnostics | | | | | Thailand |
| Increasing vegetables/ fruits | Pests/ diseases | Tomato/ potato | BW biocontrol agent | | Potato (Vietnam) | BW-resistant varieties | | | AVRDC Vietnam |
| | | Papaya | | | | | Papaya ringspot virus | | ISAAA Asian regional network |

Notes: 1–7: see Table 1.1.
FAO, Food and Agriculture Organization; IPM, Integrated Pest Management; ARBN, Asian Rice Biotechnology Network; IRRI, International Rice Research Institute; ICRISAT, International Crops Research Institute for the Semi-Arid Tropics; CIMMYT, Centro Internacional de Mejoramients de Maiz y Trigo; FMD, foot and mouth disease; AVRDC, Asian Vegetable Research and Development Center; BW, bacterial wilt; ISAAA, International Service for the Acquisition of Agribiotech Applications.

**Table 1.3.** Illustrative applications of biotechnology to the goal of environmental protection.

| | | | Biotechnology applications (1–6) | | | | | | |
|---|---|---|---|---|---|---|---|---|---|
| Environmental objective | Constraint | Target | 1 Microbial fermentation/ biocontrol/bio-fertilizers | 2 New diag-nostics/ vaccines | 3 Tissue culture/ microprop-agation | 4 Molecular markers and MAS | 5 Genetic engineering/ transgenics | 6 Genomics | 7 Examples |
| Pesticide Reduction | Pesticide misuse | Cotton | | | | | Bt cotton | | China |
| | | Rice | | | | Insect resistance | Bt genes | | IRRI |
| | | Vegetables | Bt as bio-pesticide | | | | | | Malaysia |
| Efficient water use | Drought/ salinity | Maize, rice, sorghum | | | | Molecular markers | | Cereal genes for drought tolerance | China CIMMYT/ IRRI/ICRISAT |
| Reduce deforestation | | | | Identify diversity sources | Rapidly growing fuel wood | | | | CIFOR/ ICRAF/ IPGRI |

Notes 1–7: see Table 1.1.

Bt, *Bacillus thuringiensis*; IRRI, International Rice Research Institute; CIMMYT, Centro Internacional de Mejoramiento de Maiz y Trigo; ICRISAT, International Crops Research Institute for the Semi-Arid Tropics; CIFOR, Centre for International Forestry Research (Indonesia); ICRAF, International Centre for Research in Agroforestry; IPGRI, International Plant Genetic Resources Institute (Rome).

**Table 1.4.** Illustrative applications of biotechnology to the goal of increasing competitiveness.

| | | | Biotechnology applications (1–6) | | | | | | |
|---|---|---|---|---|---|---|---|---|---|
| | | | 1 | 2 | 3 | 4 | 5 | 6 | 7 |
| Objective | Constraint | Target | Microbial fermentation/ biocontrol/ biofertilizers | New diagnostics/ vaccines | Tissue culture (tc)/ microprop-agation | Molecular markers and MAS | Genetic engineering/ transgenics | Genomics | Examples |
| Sustainable productivity exports | Yield/quality | Coconut | | | | High lauric acid | | | Philippines |
| | | Banana | | | Banana tc | | BBTV resistance | | India China Vietnam |
| Food safety and quality control | Pesticide residues | Food exports | | Certify export quality | | | | | |
| Value-added exports | Product quality | Bamboo | | | Bamboo tc | | | | China Vietnam |

Notes 1–7: see Table 1.1.
BBTV, banana bunchy-top virus.

enable the development of new products. These products may include more nutritious food with higher levels of vitamins and minerals, crop varieties with improved tolerance to pests and diseases that require less pesticide use, and vaccines that protect livestock against lethal diseases.

Because modern biotechnology provides powerful new techniques and the number and variety of new products is so great, it is important to provide appropriate regulatory mechanisms to ensure that products produced by the use of new techniques are as safe as the products of traditional biotechnology. This is especially so when these products are LMOs that might interact with the environment. The applications of modern biotechnology to agriculture, particularly the development of transgenic crops and other LMOs, are the subject of public debate as to the safety and efficacy of the new products.

A number of applications of biotechnology are being used in Asia, Africa and Latin America to address specific problems in crop and livestock production, fisheries and forestry. There are also opportunities for their expanded use to contribute to poverty reduction, food security, environmental conservation and trade competitiveness (Tables 1.1–1.4). The range of applications includes new biocontrol agents for pest and weed control, improved diagnostics and vaccines for fish and livestock diseases, higher-quality and disease-free planting material (e.g. banana, potato, sweet potato), new varieties of rice and maize selected using molecular markers for drought and salinity tolerance, and transgenic varieties of selected crops.

The most commercially widespread of the new transgenic crops are cotton varieties containing one or more genes from the bacterium *Bacillus thuringiensis* (Bt) for insect resistance. Insect-resistant, transgenic cotton varieties were grown by some 3 million small-scale farmers on at least 0.3 million ha in China in 2000.

The applications of the new developments in biotechnology could contribute more to strategies to achieve future poverty reduction and food security if they are better targeted on issues that affect poverty and food security and if they are also accompanied by political will, appropriate public policies, public and private investments in technology development and product delivery and, importantly, regulatory frameworks that generate both consumer and commercial confidence (Persley and Doyle, 1999).

## Agricultural Biotechnology in Selected Countries

### Asia/Pacific

China, India, Indonesia, Pakistan, the Philippines and Thailand are committed to the use of modern biotechnology in agriculture and are investing significant human and financial resources in this policy and have done so over the past decade.

*China* sees the greatest challenge as the use of biotechnology to increase food production and improve product quality in an environmentally sustainable manner. China has moved quickly to develop new biotechnologies. Over 103 genes have been evaluated for improving traits in 47 plant species. The crops include rice, wheat, maize, cotton, tomato, pepper, potato, cucumber, papaya and tobacco. A variety of traits were targeted for improvement, including disease resistance, pest resistance, herbicide resistance and quality improvement. Approximately 50 genetically improved organisms (GIOs) have been approved for commercial production, environmental release or small-scale field testing in China. In a few cases, new genetically improved varieties have been approved for large-scale commercial production. Genetically improved plant varieties, mainly of insect-resistant cotton, were grown commercially by some 3 million farmers on at least 0.5 million ha of land in China in 2000 (Zhang, 2000).

*India* has devoted considerable public resources to infrastructure and human resource development in biotechnology. Current efforts are aimed at applications to improve agricultural productivity; bioremediation in the environment; medical biotechnology for the production of new vaccines, diagnostics and drugs; industrial biotechnology; and bioinformatics (Sharma, 2000; see also Sharma, this volume). R & D priorities in agriculture include new regeneration protocols for rapid multiplication of citrus, coffee, mangrove, vanilla and cardamom. Yield of cardamom has increased 40% using tissue-cultured plants.

Over the past two decades, *Indonesia* has placed high priority on biotechnology. The government designated three national biotechnology centres to coordinate R & D in agriculture, medicine and industrial microbiology. Applications of biotechnology in agriculture are primarily the responsibility of the Agency for Agricultural Research and Development (AARD). A national committee on biotechnology advises the minister in developing guidelines for government policy in the promotion of biotechnology. In recent years there has been an extensive training programme within Indonesia and abroad to upgrade skills of scientists involved in biotechnological research.

Crop improvement efforts using modern biotechnology started in *Pakistan* in 1985, when a training course was held on recombinant DNA. Work is now concentrated on chickpea, rice and cotton, where there is some private-sector investment. Field evaluation of LMOs is hampered by lack of biosafety regulations. The government controls field testing, multiplication, distribution and biosafety issues. The country lacks policy and regulations regarding IPR and patents involving biotechnology (see Zafar, this volume).

*The Philippines* began their biotechnology programmes in 1980 with the creation of the National Institutes of Molecular Biology and Biotechnology, with a focus on agricultural biotechnology. In 1997, the Agriculture Fisheries Modernization Act recognized biotechnology as a major means of increasing agricultural productivity. The Act provided for a budget for agricultural biotechnology of almost US$20 million annually for the next 7 years (4% of the total R & D budget), an increase from US$1 million per year. In 1998, five high-level biotechnology research projects were funded by government: development of new varieties of banana resistant to banana bunchy-top virus and papaya resistant to ringspot virus; delayed ripening of papaya and mango; insect-resistant maize; marker-assisted breeding in coconut; and coconut with high lauric acid content. Public concerns about the safety of LMOs have been vocal in the Philippines and this is constraining the commercial use of modern biotechnology in agriculture (de la Cruz, 2000).

*Thailand* is focusing on the applications of biotechnology to traditional foods, fruits and export commodities. R & D priorities are to raise production and cut costs by using new biotechnology to address problems on crops such as rice, sugar cane, rubber, durian and orchids. An early success in Thailand has been in the application of biotechnology to the development of new molecular diagnostics for the diagnosis and control of virus diseases in shrimps. These diseases cost the shrimp export industry over US$500 million in lost production in 1996 (Morakot, 2000).

Five of these countries have regulatory systems in place at the national and institutional level to govern R & D programmes and commercial developments where appropriate. Intellectual property management was considered to be a difficult issue for all six countries.

## Latin America and the Caribbean

The main challenges in relation to agricultural biotechnology are: management of intellectual property for both major and minor crops; assessment of several research options, not only a molecular approach, in assessing how best to tackle problems and challenges to improve agricultural productivity; identification of beneficiaries; prioritization of work on favoured and/or marginal areas; use of GIOs as indicators of environmental damage; and need to monitor the behaviour of GIOs in the environment after release. The ecological research effort for monitoring GIOs is needed to satisfy public concerns about the behaviour of GIOs in the environment, and needs to focus on the following key questions. What are the specific concerns? How to do it? Who will do it? Who will pay for it?

In *Brazil*, many lines of R & D are benefiting from the application of biotechnology tools such as marker-assisted plant and animal breeding, genomic mapping of several species including sugar cane, embryo transfer applied to different animal species, genetic resources characterization and conservation, and use of genetic improvement to introduce new traits, such as papaya resistant to papaya ringspot virus and beans resistant to golden mosaic virus. The issues of field testing of genetically improved plants need to be addressed. Tropical agriculture is very different from the temperate fields where most of the new genetically improved products have been tested. Protocols are required for field trials, risk assessment (environmental and food safety), registration of products and public acceptance. The need is urgent, because these are constraints that will intensify as GIOs become an integral part of the research agenda in the region (Sampaio, 2000).

In *Costa Rica*, there is particular interest in using the tools of biotechnology to characterize and conserve biodiversity (Sittenfeld *et al.*, 2000). Costa Rican institutions have developed some innovative partnerships for bioprospecting, which could serve as a model for other countries. In agriculture, pesticide use increased threefold between 1993 and 1996 on crops such as banana, coffee and rice. Much of this pesticide is used to control banana diseases, where pesticide use is a health risk to field workers and an environmental risk to land, water and animals. Biotechnology-based solutions are urgently needed to replace chemical control of banana diseases. Recent discoveries in regard to understanding and managing the development of fungicide resistance in the black Sigatoka pathogen are leading to a 10–15% reduction in fungicide use on banana (a potential saving of some US$10 million per year). On rice, new virus-resistance genes are being introduced into local rice varieties.

Coffee continues to be the major export crop for *Colombia*, though its importance has declined as exports of other commodities (e.g. banana, sugar, beef, cut flowers) have increased. Agriculture now accounts for 19% (estimated in 1999) of Colombia's gross domestic product (GDP). There are extensive national efforts in many areas of biotechnology, which are aimed at delivering benefits to the rural areas while also maintaining a healthy environment and sustainable national agriculture.

## Africa

The major challenge for *Kenya*, *Zimbabwe* and *Egypt* is the persistent poor performance of agriculture in Africa generally, which is leading to a food crisis. The issues concerning many countries are how to improve food security, increase productivity, conserve biodiversity, reduce pest management costs and deal with increasing urban migration. Specific issues related to biotechnology are how to develop institutional capacity for risk assessment and management, to access information on developments in biotechnology elsewhere that may have application in Africa and to develop the necessary human resources and infrastructure.

Several success stories are coming out of Africa, where biotechnological approaches have contributed to the solution of specific problems, reduced the cost of pest control and created new employment opportunities in towns and villages. They include the wide adoption by farmers of rapid multiplication of disease-free banana plantlets in Kenya; use of new genetically improved pest-resistant cotton by small farmers in South Africa; and use of new vaccines against animal diseases in Kenya and Zimbabwe (Chetsanga, 2000; Ndiritu, 2000; Njobe-Mbuli, 2000).

Some of the problems and constraints identified include: lack of awareness of the benefits and risks associated with modern biotechnology; lack of capacity in some countries to deal with assessing these benefits and risks and in regulating the use of modern biotechnology; high investment costs associated with biotechnological innovations; and increasing concerns being expressed in the media about the potential negative impacts of biotechnology and the need for public awareness of the issues.

In sub-Saharan Africa the need is both to improve awareness and institutional capacity to develop biotechnology-based products and, perhaps as importantly, for African stakeholders, scientists and policy-makers to articulate an African agenda and to participate in critical global debates on trade and economic growth (Njobe-Mbuli, 2000).

One of the major targets for biotechnology in *Egypt* is the production of transgenic plants conferring resistance to the biotic and abiotic stresses that are causing serious yield losses in many economically important crops in the country. The Agricultural Genetic Engineering Research Institute (AGERI) was established in 1990 with the aim of mobilizing the most recent technologies available worldwide to address problems facing agricultural development (Madkour, 2000).

The challenges identified were the need to increase agricultural productivity, while preserving the fragile natural resource base in the region and the need to conserve the rich indigenous plant and animal species. The opportunities include: using modern biotechnology to develop crop varieties tolerant to biotic and abiotic stresses, especially drought and salt tolerance; improving the nutritional quality of agricultural commodities; producing biofertilizers and biopesticides; and improving the availability of soil nutrients.

The main constraints are inadequate financial resources, lack of qualified personnel, poor infrastructure and insufficient regional and international collaboration. There is also a lack of clear strategies, policies and regulatory frameworks to guide the use of modern biotechnology in most countries of the region.

## Risks and Benefits in the Use of LMOs

Public concerns about the use of LMOs lie in four major areas: food safety, the environment, socio-economic and ethical issues. First, in relation to food safety, there are concerns about assessing the risks of genetically modified (GM) foods to human health and understanding potential benefits of new GM foods to consumers. Secondly, in relation to environmental effects, the concerns relate to assessing the risks and benefits of releasing LMOs into the environment and the effects such releases may have on the environment. These may be direct effects, such as on biodiversity, or indirect, through changing agricultural practices that affect the environment. LMOs released for agricultural purposes may be plants, trees, livestock, fish and/or microorganisms.

Given the rapid pace of new developments in agricultural biotechnology, many consumers are seeking further information about the potential effects of biotechnology on their food and their environment. The media have sensed that this is an issue high in public consciousness and are actively promoting widespread debate on how we should best use the discoveries of modern genetics. This is one of the most important public debates of the new millennium, because its resolution will have global implications for food and raw material production for the rest of this century (Johnson, 2000).

## Food safety and human health

*Risk factors.* There are several areas of public concern with regard to potential human health risks of GM foods. These relate to understanding the potential of proteins and/or other molecules in GM foods to cause allergic reactions, to act as toxins or carcinogens and/or to cause food-intolerance reactions among the population. Methods of testing and evaluating these types of risks have been established for food and these are being applied to GM foods so as to detect any increased risks associated with particular foods (Lehrer, 2000).

A recent consultation between the Food and Agriculture Organization (FAO) and the World Health Organization (WHO) reported that 'the Consultation was satisfied with the approach used to assess the safety of the genetically modified foods that have been approved for commercial use'. The US Food and Drug Administration (FDA) has also stated that 'we have seen no evidence that the bioengineered foods now on the market pose any human health concerns or that they are in any way less safe than crops produced through traditional breeding'.

Although no instances of harmful effects on human health have been documented from GM foods, that does not mean that risks do not exist as new foods are developed with novel characteristics. GM foods should be assessed on a case-by-case basis, using scientifically robust techniques, so as to ensure that the foods brought to market are safe for human consumption.

For example, any protein added to a food should be assessed for its potential allergenicity, whether it is added by genetic engineering or by manufacturing processes (Lehrer, 2000). Allergenicity can be raised in foods either by raising the level of a naturally occurring allergen (e.g. in groundnuts) or by introducing a new allergen. More than 90% of the food allergens that occur in 2% of adults and 4–6% of children are associated with eight food groups. These include Crustacea, eggs, fish, groundnuts, soybean, tree nuts and wheat. These foods merit close attention when examining GM foods for the potential for increased risk of allergenicity (Lehrer, 2000). There is also a need to assess the allergenic potential of unknown proteins, such as those produced by Bt genes in plants. It was the presence of a heat-tolerant Bt protein in Starlink maize that caused the US FDA to withhold approval for its use in human consumption, as the FDA scientific advisory panel considered that this protein posed a moderate allergy risk.

*Antibiotic resistance.* There are also concerns about the risk that the antibiotic-resistance genes used as selectable markers in GM plants may transfer to microorganisms that are human pathogens, adding to the increasing problem of antibiotic resistance in human pathogens. This problem is the result of widespread use of antibiotics in human and animal health. WHO, the Organization for Economic Cooperation and Development (OECD) and FAO have assessed the antibiotic-resistance markers used in transgenic crops as being safe. The risk of transfer of an antibiotic marker from a GM food to a human pathogen is considered remote. Nevertheless, the use of these antibiotic markers is being phased out. Other selectable markers that can be removed from the plant in the development phase are replacing them (Escalar, 2001).

*Risk assessment*

The International Life Sciences Institute (ILSI) has developed a decision tree that provides a framework for risk assessment in foods (Lehrer, 2000). It uses the following criteria, that an introduced protein in a food is not a concern if: (i) there is no history of common allergenicity; (ii) there is no amino acid sequence similar to those of known allergens; (iii) there is rapid digestion of the protein; and (iv) the protein is expressed at low levels. For example, these risk assessment techniques were used to test the safety of increasing the protein content in soybean by introducing a protein from Brazil nut. However, food allergy tests showed that this transferred a potential allergen to soybean. Hence, further development of this GM high-protein soybean ceased.

The techniques for assessing the potential for allergenicity, toxicity and carcinogens in food are well established and should be readily able to be used by trained professionals in many countries (Metcalf *et al.*, 1996; Lehrer, 2000). Given increasing global concerns about food safety, all countries will need to have in place food-safety regulations and the human and institutional capacity to be able to ensure the safety of their food supply (World Bank, 2000).

*Benefits to human health*

The risks in GM foods need to be weighed against the benefits. The next generation of GM foods is likely to include a number of functional foods that offer some nutritional benefits to consumers. Human health benefits of GM foods lie in the potential for introducing traits that enable factors such as the following:

- Improved nutritional quality of foods (e.g. higher vitamin content, lower fat content).
- Reduced toxic compounds in food (e.g. cassava with lower levels of cyanide).

- Crops grown with lower levels of chemical pesticides, thus reducing pesticide residues in food.

*Labelling*

A key concern of consumers is being able to identify those foods that may contain allergens and other potentially harmful substances. Those who have allergic or food-intolerant reactions to particular foods can avoid them. Others may wish to avoid certain foods on health, ethical or religious grounds. Informative food labelling, whether mandatory or voluntary, could be used to provide information about specific products and enable consumers to make informed decisions about their use, in terms of both risks and potential beneficial effects (Skerritt, 2000).

*International developments*

As a result of an international conference on the safety of GM foods, the OECD (2000) noted the need for the following:

- Factual points of departure as to where there is agreement and disagreement on human health risks.
- Benefits versus risks, which differ for different countries and environments.
- Management of genetic modification technologies.
- The role of stakeholders and consultative processes.
- An international programme of activities to inform the public debate and policy-making, including a possible international panel to review scientific evidence on the benefits and risks of the applications of new biotechnologies in food.

The International Council of Scientific Unions (ICSU) has initiated a review of the scientific basis of assessing risks and benefits of GM foods, as a contribution to the ongoing debates, nationally and internationally.

## Environmental risks and benefits of LMOs

In regard to the risks and benefits of LMOs in the environment, public concerns are based on the premise that, when LMOs contain genes introduced from outside their normal range of sexual compatibility, these organisms may present new risks to the environment. Present recombinant DNA technology enables such genetic modifications to be made to introduce new and potentially useful traits into plants, trees, microorganisms, livestock and fish.

The most widespread new LMOs released in the environment are GM crops, with some 44 million ha being cultivated commercially, in 15 countries. Most of this area (68%) is planted in North America with GM maize, cotton, potato, rape-seed and soybean, modified with new genes for insect resistance and/or herbicide tolerance and virus resistance (James, 2000). Among emerging economies, China has at least 0.5 million ha, mainly of transgenic cotton with insect resistance, being grown by some 3 million farmers.

The concerns about the impact of LMOs on the environment are about the risks of both direct ecological effects and indirect environmental effects due to changing agricultural management practices brought about by the use of LMOs. The applications of biotechnology may offer means that enable agriculture to sustain yields while minimizing the adverse environmental effects of agricultural intensification. Yet there is a perception that some of the present generation of GM crops, especially the new herbicide-tolerant and insect-resistant crops developed for extensive agricultural systems, may present yet further risks to biodiversity in present agricultural systems (Johnson, 2000).

Some of the claims for the potentially harmful effects of currently cultivated GM crops on the environment include the following:

- The use of genes from Bt as a source of resistance to insect pests may lead to 'super' pests.
- The use of crops with resistance to glyphosate (Round-Up) may lead to greater use of broad-spectrum herbicides.
- Virus-derived genes used as a source of virus resistance in crop plants may lead to new viruses with potential to kill native plants.

Governments, research organizations and companies must respond to these claims and other concerns and have in place the means to scientifically assess and report on the risks and the benefits to the environment of LMOs (Cook, 2000).

There is a need for science-based risk assessments for plants and other organisms (livestock, fish, trees and microbes) that are intended for use in agricultural and other managed environments. There is also an urgent need for ecological research and developing agreed standards and protocols to enable the continuing monitoring of the behaviour of LMOs after experimental (small-scale) and commercial (large-scale) releases into the environment. Such data would then feed back into risk assessments, so as to inform future decisions on the development and management of LMOs being developed for agricultural purposes.

A recent review of the scientific literature reveals that key experiments on both the environmental risks and benefits of genetically engineered plants are lacking. (Wolfenberger and Phifer, 2000). The complexity of ecological systems presents considerable challenges to designing experiments to assess such risk and benefits.

The US National Academy of Sciences (NAS, 1987) released one of the earliest studies on the safety of LMOs in the environment. Its four conclusions were as follows:

- There is no evidence that unique hazards exist either in the use of recombinant DNA techniques or in the transfer of genes between unrelated organisms.

- The risks associated with the introduction of recombinant DNA-engineered organisms are the same in kind as those associated with the introduction into the environment of unmodified organisms and organisms modified by other genetic techniques.
- Assessment of the risks of introducing recombinant DNA-engineered organisms into the environment should be based on the nature of the organism and the environment into which it will be introduced (product), not on the method (process) by which it was modified.
- There is an urgent (and ongoing) need for the scientific community to provide guidance for both investigators and regulators in evaluating planned introductions of modified organisms from an ecological perspective.

Thousands of field trials conducted with GM crops over the past decade support these conclusions by NAS in this and subsequent reports (NAS, 1987; NRC, 1989, 2000; Cook, 2000).

*GM plants in the environment*

Cook (2000) reported on an approach to science-based risk assessment for plants intended for use in agricultural or other managed environments. In addressing the risks posed by the cultivation of plants in the environment, five environmentally related safety issues need to be considered (OECD, 1993). These issues are the potential for:

- Gene transfer, meaning the movement of genes from a crop through outcrossing with wild relatives to form new hybrid plants.
- Weediness, meaning the tendency of a plant to spread beyond the field where first planted and establish itself as a weed or invasive species.
- Trait effects, meaning effects of traits that are potentially harmful to non-target organisms.
- Genetic and phenotypic variability, meaning the tendency of the plant to exhibit unexpected characteristics.
- Expression of genetic material from pathogens, such as the risk of genetic recombinations following mixed virus infections.

*Gene flow and transfer of traits to other species.* Gene transfer is an issue when crops are being grown in areas close to their wild relatives, with whom they are able to cross to form interspecific hybrids. Natural hybridization occurs within 12 of the world's 13 most important food crops and their wild relatives (the exception being banana, since cultivated banana is infertile). Thus, the world's major cereal crops (maize, wheat, barley, sorghum), oil-seeds (rape-seed, soybean and groundnut) and root and tuber crops (cassava and potato) can cross with their compatible wild relatives. Such wild relatives occur in the centres of diversity of these crops (see map of the centres of origin and diversity of the world's major food crops in Serageldin and Persley (2000)). Natural hybridization may occur at low frequency when pollen blows or is otherwise transported from crops to wild relatives in the vicinity. Such gene flow and interspecific crosses cannot occur in crops whose centres of origin and diversity and closest wild relatives are on other continents (e.g. maize in Europe, since its centre of origin is in Mexico).

Recent research confirms that genes introduced into some GM crops may spread into related native species (Chevre *et al.*, 1997). This is not unexpected since genes have long been known to move from conventionally bred crops to wild relatives, at low frequency. For example, in the UK, such hybrids occasionally occur between oil-seed rape (*Brassica napus*) and native species, such as wild turnip (*Brassica rapa*). Published studies on the gene transfer issue are dominated by rape-seed (Wolfenberger and Phifer, 2000), whose centre of origin is in Europe.

The difference is that genes inserted into GM crops are often derived from other phyla, giving traits that have not been present in wild plant populations. The concern is that these genes may change the fitness and population dynamics of hybrids formed between native plants and related GM crops, eventually back-crossing genes into the native species. The importance of pollen transfer from GM crops to wild relatives is not that it occurs but whether the resulting hybrids survive and reproduce and introgress genes back into the native population.

The issue is not so much the rate of gene flow (on which there has been much research), as the impact that this might have on agriculture and the environment (on which there has been very little research). Conventional plant breeding, using techniques such as mutagenesis and embryo rescue, also produces new genes in crops, about which we also know very little regarding their behaviour in the wild (Johnson, 2000).

*Weediness.* There are risks that GM plants could have negative impacts on natural ecosystems by increasing weediness by two routes. First, the GM plants could establish self-sustaining populations outside cultivation themselves. The concern is that these plants may become invasive weeds that out-compete wild populations and thus lead to further decreases in biodiversity in native plant habitats. Weeds having tolerance to a range of herbicides could also emerge (Johnson, 2000).

Secondly, novel genes from GM crops could be introduced into their wild relatives by pollen spread and the survival and reproduction of the resulting hybrids. This may have a negative impact on the wild plant population if new genes are introgressed back into the wild plant population. For this to happen, the new genes must increase the plants' fitness to survive and reproduce in the wild.

Transfer of certain genes, such as resistance to insects, fungi and viruses, may increase the fitness (ability to reproduce) of any resulting hybrids. If hybrids acquired insect resistance from GM crops, they could damage food chains dependent on insects feeding on previously non-toxic wild plants. Not only would there be a direct effect, for many insects are entirely dependent on single plant species, but acquisition of resistance in wild plants would probably change their population dynamics, increasing the risks of their invading agricultural land and natural ecosystems.

Many geneticists would argue that most 'foreign' genes introduced accidentally from GM crops to crop/native plant hybrids would decrease their fitness in the wild, leading to rapid selection of these genes out of the population (Johnson, 2000). A recently published 10-year study by Imperial College in the UK on the fitness of GM plants to survive in the wild supports this hypothesis. GM maize, rape-seed and sugar beet (all with herbicide tolerance) and potato with insect resistance were compared with conventionally bred crops. All four GM crops were out-competed by their conventionally bred relatives (Crawley *et al.*, 2001). Thus, in this experiment, the genetic modifications in these species for herbicide tolerance and insect resistance made them less competitive and less fit to survive in the wild. Plant breeding tends to reduce rather than increase the weediness characteristics of crop plants (Cook, 2000).

*Trait effects.* Trait effects are those that are harmful to non-target organisms. For example, plants modified to produce pesticidal proteins, such as Bt toxins, may have both direct and indirect effects on populations of non-target species. One group of Bt toxins primarily targets Lepidoptera (butterflies and moths, particularly the European corn-borer) and the others affect Coleoptera (beetles). The effects of Bt toxin-producing plants on non-pest species among these insect groups may vary widely, depending on sensitivity and the concentration of Bt in transgenic plants and environmental conditions.

Laboratory experiments demonstrated that the larvae of monarch butterflies (a relative of the European corn-borer) were susceptible to pollen from Bt maize when ingested in large amounts (Losey *et al.*, 1999), but the actual ecological significance of this laboratory experiment was not clear. Subsequent field experiments in several locations in North America found that there were no significant differences between butterfly survival in areas planted with Bt maize and those planted with conventional crops (Henderson, 2000).

In assessing trait effects, such as those of Bt crops, on non-target species, it is important to compare the potential risks of Bt crops with the present effects of chemical pesticide use in risk assessments.

*Genetic and phenotypic variability.* This is the tendency of a plant to exhibit unexpected (pleotropic) characteristics in addition to the expected characteristics. This trait is well known from conventional breeding, but becomes an identifiable hazard only if the variability leads to one of the other safety issues, such as greater weediness or a greater tendency for out-crossing.

*Expression of genetic material from pathogens.* Another potential hazard would be the possibility of recombination of a virus gene expressed by the plant with genes from another virus infecting that plant. This risk would be similar to the risk of genetic recombinations following mixed virus infections that occur in nature.

*Genetic modification of native species.* There is some experimentation on genetic modification of native species (e.g. *Eucalyptus* in Australia). These developments greatly increase the risks of gene transfer because GM native organisms will be completely cross-fertile with native species. There is also a risk that GM varieties of native plants would be fitter than native species and colonize natural ecosystems, with unpredictable results (Johnson, 2000).

*Biotechnology and biodiversity.* Risks to biodiversity and wildlife are important issues in particular environments. Careful assessment is necessary of the risks associated with the creation of new selection pressures coming from the introduction of GIOs into the environment. These new selection pressures may have profound effects on the delicate balance of life. Of special concern is the potential impact on biodiversity of GIOs as selection pressures wield influence in the species composition of the ecosystem. These concerns merit further study, especially on the behaviour of GIOs in the open environment. The framework for strategic planning in the deployment of GIOs should be formulated with sustainability as the primary concern (Johnson, 2000; Platais and Persley, 2001).

*Ecological research.* There are concerns that the science needed to be able to assess these ecological risks is not being undertaken (Johnson, 2000). At the moment we do not know what effect escaped genes might have on natural and farmland ecosystems. This lack of science is disturbing, given the commercial pressure for the introduction of GM crops into the landscape (Johnson, 2000). There is clearly a need to set up effective monitoring systems to detect gene transfer and research to assess its ecological impacts. Research in this area would be in the interests of both the industry and the environment as it would yield data that would form a scientific basis for regulatory decisions.

There may also be scientific options that could be used in future generations of GM crops that would mitigate some of the ecological risks. For example, it may be possible to include in GM crop plants inbuilt mechanisms, such as pollen incompatibility, to prevent gene flow. Another means to ensure genetic isolation is to make sure that, wherever possible, plants used for transformations are unrelated to native species and edible crops within the intended market territory. This principle would influence the choice of which plants companies choose as platforms for biomedical and industrial product transformations (e.g. higher starch production, vaccine production in plants). If gene technology is to become a standard technique for plant breeding, genetic isolation of crops from the rest of the living environment may become normal practice, as will the removal of marker genes for antibiotic resistance (Johnson, 2000).

### *GM crops and agricultural intensification*

The prospect of gene transfer causes concern for crops that have wild relatives in the same geographical area. Perhaps of greater importance is the fact that management of some GM crops would be very different from conventional intensive agriculture or organic farming.

In the USA, GM herbicide-tolerant (GMHT) crops are grown under a regime of broad-spectrum herbicides applied during the growing season. Farmers report almost total weed elimination from GMHT crops, which include cotton, soybean, maize, sugar beet and oil-seed rape. Recent research in the UK confirms that weed control in GM beets and other GMHT crops is likely to become much more efficient. These results are hardly surprising, since this is the main purpose behind the technology.

This GMHT system will soon be available, at least experimentally, for many agricultural crops, including vegetables. Broad-spectrum herbicides used on commercial scale GMHT crops during the growing season may be far more damaging to farmland ecosystems than the selective herbicides they might replace. Using these herbicides in the growing season may also increase the impact of spray drift on to marginal habitats and watercourses. It is not the volume of herbicides that is the issue but their efficiency and impact on wildlife. When insect resistance and herbicide tolerance are combined in the same crop variety, there may be few insects capable of feeding on the crops and few invertebrates and birds would be able to exploit the weed-free fields (Johnson, 2000).

The use of more effective pesticides (including herbicides) over the past 20 years has been a major factor in causing serious declines in farmland birds, arable wild plants and insects in several European countries. Pesticides not only have direct toxic effects on wildlife but they also enable modern crop-management changes to take place.

Besides the aesthetic and scientific reasons for conserving biodiversity within and around agricultural crops, there is another important utilitarian reason for wanting to do so. There is a danger of losing the food-chain links between native species and crop systems. This link is vital for preserving the function of biodiversity to deliver early warning of dangers in crops or the chemicals used to manage them.

There is some evidence that the use of GM crops with insect resistance in North America is reducing the volume and frequency of pesticide use on cotton, maize and soybean (Wolfenberger and Phifer, 2000).

In addition, the future development of new crops with improved tolerance to abiotic factors (such as drought, salinity, frost) and the advent of 'pharmed' crops used to produce vaccines and industrial products may also change crop-management practices. They may either increase or decrease demand for arable land in the long term. They may also put further pressure on natural biodiversity on marginal land.

The problem with assessing the environmental impact of these changes in management is that the regulatory system and the public have little scientific data on which to assess the real risks and potential benefits from adopting GMHT crop systems. In the UK, 27 field-scale experiments are in place to try to answer some of these important questions (Johnson, 2000). Research is urgently required to make the ecological consequences of using GM crops clearer. Information from such research can then be used by regulators to make more informed and publicly defensible decisions about whether GM crops should be commercialized and under what conditions.

*Future R&D strategy*

There are some promising new developments in R & D that may assist in the design of future GM crops that would have clear benefits to the environment and that would mitigate some of the environmentally damaging effects of agricultural intensification (Johnson, 2000).

- Securing fungal resistance in adult plants by 'switching on' resistance genes that are active in the seed, but not currently in adult plants. This seems to be an elegant and safe use of biotechnology that could lead to significant reductions in fungicide use.
- Achieving insect resistance by altering physical characteristics of plants, perhaps by increasing hairiness or thickening the plant cuticle. This could reduce insecticide use, without using in-plant toxins.
- Altering the growing characteristics of crops (for example, shortening the growing season or changing the timing of harvests) offers the prospect of allowing more fallow land and less autumn planting. The recent discovery of dwarfing genes could be a significant step towards the production of higher-yielding and more reliable spring-sown cereals.
- Developing crops (including trees) that can tolerate high levels of natural herbivory and yet remain viable.
- Preventing out-crossing by engineering pollen incompatibility and other mechanisms into crops. This could significantly reduce the risk of spread of GM traits into native species.

Many of these traits could simply be transferred from one crop variety into another or be accomplished by switching on or off genes already present in the plant. Such transformations are likely to be more acceptable to the public than moving genes between phyla. The consequences of short-circuiting genetic distance between species, which has been maintained over long periods of time and geographical isolation, are not yet well enough understood to be able to assess the risks (Johnson, 2000).

Biotechnology and the new science of genomics, which is giving new insights into how genes function, offer a whole new range of options for how to use land. For the first time, it may be possible to design crops to suit the land and the purpose rather than having to adapt land and purpose to suit the crop. These could become important components of sustainable farming systems that combine yield increases with environmental sustainability. This is also important for developing countries where biotechnology may be able to offer new solutions to old problems of crop pests and disease in locally adapted crops, rather than trying to export conventional, chemically based agriculture with its damaging effects on biodiversity and the wider environment.

### Regulatory systems

Food safety and biosafety regulations should reflect international agreements and a given society's acceptable risk levels, including the risks associated with not using biotechnology to achieve desired goals.

The principles and practices for assessing the risks on a case-by-case basis are well established in most OECD countries and several emerging economies. These principles and practices have been summarized in a series of OECD reports published over the past decade or more. National, regional and international guidelines for risk assessment and risk management provide a basis for national regulatory systems. Biosafety guidelines are available from several international organizations, including the OECD, United Nations Environment Programme, United Nations Industrial Development Organization and the World Bank.

Regulatory trends to govern the safe use of biotechnology, to date, include undertaking scientifically based, case-by-case hazard identification and risk assessments; regulating the end-product rather than the production process itself; developing a regulatory framework that builds on existing institutions rather than establishing new ones; and building in flexibility to reduce regulation of products after they have been demonstrated to be of low risk.

All sections of society should be included directly in the debate and decision-making about technological change, the risks of that change and the consequences of no change or alternative kinds of change.

## Ethical Issues

In regard to ethical issues, it is important to pursue a dialogue on ethical issues to clarify moral and ethical issues of concern and how they might be addressed in different societies. The ethical challenges include the role of science – its risks, benefits and impact on society. Moral and ethical standards are used to develop laws governing some aspects of biotechnology (e.g. in medicine, laws governing human cloning).

A major ethical concern is that 'genetic engineering' and 'life patents' accelerate the reduction of plants, animals and microorganisms to 'mere commercial commodities, bereft of any sacred character'. All agricultural activities constitute human intervention into natural systems and processes, and all efforts to improve crops and livestock involve a degree of genetic manipulation. Continued human survival depends on precisely such interventions.

## Intellectual Property Management

Many R & D programmes face the challenge and opportunities of managing intellectual property. Partnerships are critical to effective management and investment in intellectual property protection.

- Learning to manage IPR is a critical issue for many countries and institutions.
- Intellectual property management includes clarifying the role of institutions, developing an inventory of intellectual property, developing ownership of intellectual property where appropriate, undertaking technology transfer and marketing the intellectual property.
- Human resource development is a major need in this area.
- Benefit sharing with holders of indigenous knowledge and genetic resources is an important issue that must be addressed.

It is most important to build up human resource capacity in IPR for scientists, managers, policy-makers and society as a whole. Societal changes are reflected in changing IPR requirements and further changes are likely to result from continuing international negotiations on IPR and finding ways to reflect the contribution of indigenous knowledge.

### Public/private-sector roles

In order to maximize the use of modern molecular knowledge, both public- and private-sector research is required to bring innovation and choices to farmers and consumers. The private sector is likely to focus on those areas of opportunity that will repay their investment in innovation. The public sector must maintain the freedom to operate in an era of increasing proprietary technology. In developing countries, the public sector will need to develop technologies that meet the needs of the non-commercial sector, including the needs of resource-poor farmers and urban consumers.

## Future Actions

There is a need for more investment in public research in national agricultural research systems (NARS) and the international agricultural research centres (IARCs), to develop appropriate technologies and products that address the objectives of poverty reduction, food security, sustainable natural resources and/or trade competitiveness in developing countries. This needs to be done in partnership with the private sector (especially local companies). Farmers and consumers must be actively involved in articulating their problems and driving the R & D agenda. Partnerships and dialogues with NGOs and civil society are also needed to reach consensus as to appropriate technology choices.

There is also need for the following:

- More public and private R & D investments in targets that affect the livelihoods of the poor and that are perceived to benefit both farmers and urban consumers.
- The start-up of local companies to commercialize and distribute new technologies, including the continuing importance of local seed companies in the distribution of new plant varieties.

- Innovative mechanisms to stimulate more R & D on the problems important to the rural and urban poor, including exploring the feasibility of tax concessions in OECD countries and a global competitive grants facility.
- Exploring new modalities for public/private-sector partnerships, learning from past experience of those already in operation, especially in relation to intellectual property management.

## Conclusion

Biotechnology is only one tool, but a potentially important one, in the struggle to reduce poverty, improve food security, reduce malnutrition and enhance the living standards of the rural and urban poor. The uncertainties and the risks are yet to be fully understood and the possibilities are as yet not fully exploited. It is important not to deny people access to new technology that may address their present problems, so long as they are fully informed of the potential risks and benefits and able to make their own choices.

By assessing the current and potential usefulness of modern biotechnologies for the solution of specific problems in agriculture, new ground is being broken in analysing how best to assess and mobilize:

- Rapid developments in science and technology.
- New public policy requirements.
- New institutional arrangements.

The exchange of a wealth of knowledge, information and experience and the sharing of differing perspectives will be valuable in moving ahead with responsible dialogue and debate on the use of the new developments in science and technology for the benefit of society.

## References

Arabidopsis Genome Initiative (2000) Analysis of the genome sequence of the flowering plant *Arabidopsis thaliana*. *Nature* 408 (14 December), 796–815.

Chetsanga, C.J. (2000) Zimbabwe: exploitation of biotechnology in agricultural research. In: Persley, G.J. and Lantin, M.M. (eds) *Agricultural Biotechnology and the Poor: Proceedings of an International Conference, Washington, DC, 21–22 October 1999*. Consultative Group on International Agricultural Research, Washington, DC, pp. 118–120.

Chevre, A.-M., Eber, F., Baranger, A. and Renard, M. (1997) Gene flow from transgenic crops. *Nature* 389 (30 October), 924.

Cook, R.J. (2000) Science-based risk assessment for the approval and use of plants in agricultural and other environments. In: Persley, G.J. and Lantin, M.M. (eds) *Agricultural Biotechnology and the Poor: Proceedings of an International Conference, Washington, DC, 21–22 October 1999*. Consultative Group on International Agricultural Research, Washington, DC, pp. 123–130.

Crawley, M.J., Brown, S.L., Hails, R.S., Cohen, D.D. and Rees, M. (2001) Biotechnology: transgenic crops and natural habitats. *Nature* 6821, 682–683.

de la Cruz, R.E. (2000) Philippines: challenges, opportunities and constraints in agricultural biotechnology. In: Persley, G.J. and Lantin, M.M. (eds) *Agricultural Biotechnology and the Poor: Proceedings of an International Conference, Washington, DC, 21–22 October 1999*. Consultative Group on International Agricultural Research, Washington, DC, pp. 58–63.

Doyle, J.J. and Persley, G.J. (1996) *Enabling the Safe Use of Biotechnology: Principles and Practice*. Environmentally Sustainable Development Studies and Monographs Series No. 10, World Bank, Washington, DC.

Doyle, J.J. and Persley, G.J. (1998) New biotechnologies, an international perspective. In: *Investment Strategies for Agricultural and Natural Resources Research*. CAB International, Wallingford, UK.

Escalar, M. (2001) *The Current State of Antibiotic Resistance Marker Controversy in GM Plants*. Crop Biotechnology Knowledge Center, International Service for the Acquisition of Agri-biotech Applications, Los Baños, Philippines, 4 pp.

Henderson, M. (2000) *The Times*, London, 14 December 2000.

IFPRI (1997) *The World Food Situation: Recent Developments, Emerging Issues and Long-term Prospects*. International Food Policy Research Institute, Washington, DC.

James, C. (1999) *Global Review of Commercialized Transgenic Crops*. ISAAA Brief, International Service for the Acquisition of Agri-biotech Applications (ISAAA), Ithaca, New York.

James, C. (2000) *Global Review of Commercialized Transgenic Crops*. ISAAA Brief, International Service for the Acquisition of Agri-biotech Applications (ISAAA), Ithaca, New York.

Johnson, B. (2000) Genetically modified crops and other organisms: implications for agricultural sustainability and biodiversity. In: Persley, G.J. and Lantin, M.M. (eds) *Agricultural Biotechnology and the Poor: Proceedings of an International Conference, Washington, DC, 21–22 October 1999*. Consultative Group on International Agricultural Research, Washington, DC, pp. 131–138.

Lehrer, S.B. (2000) Potential health risks of genetically modified organisms: how can allergens be assessed and minimized? In: Persley, G.J. and Lantin, M.M. (eds) *Agricultural Biotechnology and the Poor: Proceedings of an International Conference, Washington, DC, 21–22 October 1999*. Consultative Group on International Agricultural Research, Washington, DC, pp. 149–155.

Losey, J.E., Rayor, L.S. and Carter, M.E. (1999) Transgenic pollen harms monarch larvae. *Nature* 399, 214.

McCalla, A.F. (1998) *The Challenge of Food Security in the 21st Century*, Convocation Address, Faculty of Environment Sciences, McGill University, 5 June 1998, Montreal, Quebec.

Madkour, M. (2000) Egypt: biotechnology from laboratory to the marketplace: challenges and opportunities. In: Persley, G.J. and Lantin, M.M. (eds) *Agricultural Biotechnology and the Poor: Proceedings of an International Conference, Washington, DC, 21–22 October 1999*. Consultative Group on International Agricultural Research, Washington, DC, pp. 97–99.

Metcalf, D., Astwood, J., Townsend, R., Sampson, H., Taylor, S. and Fuchs, R. (1996) Assessment of the allergenic potential of foods derived from genetically engineered crop plants. *Critical Reviews in Food Science and Nutrition* 36 (S), S165–S186.

Morakot, T. (2000) Thailand: biotechnology for farm products and agro-industries. In: Persley, G.J. and Lantin, M.M. (eds) *Agricultural Biotechnology and the Poor: Proceedings of an International Conference, Washington, DC, 21–22 October 1999*. Consultative Group on International Agricultural Research, Washington, DC, pp. 64–73.

NAS (1987) *Introduction of Recombinant DNA-engineered Organisms into the Environment: Key Issues*. National Academy of Sciences, Washington, DC, 24 pp.

Ndiritu, C.G. (2000) Kenya: biotechnology in Africa: why the controversy? In: Persley, G.J. and Lantin, M.M. (eds) *Agricultural Biotechnology and the Poor: Proceedings of an International Conference, Washington, DC, 21–22 October 1999*. Consultative Group on International Agricultural Research, Washington, DC, pp. 109–114.

Njobe-Mbuli, B. (2000) South Africa: biotechnology for innovation and development. In: Persley, G.J. and Lantin, M.M. (eds) *Agricultural Biotechnology and the Poor: Proceedings of an International Conference, Washington, DC, 21–22 October 1999*. Consultative Group on International Agricultural Research, Washington, DC, pp. 115–117.

NRC (1989) *Field Testing Genetically Modified Organisms*. National Academy of Sciences, Washington, DC, 170 pp.

NRC (2000) *Genetically Modified Pest-protected Plants: Science and Regulation*. US National Research Council, Washington, DC.

OECD (1993) *Safety Considerations for Biotechnology: Scale-up of Crop Plants*. Organization for Economic Cooperation and Development, Paris, France, 40 pp.

OECD (2000) GM food safety: facts, uncertainty and assessment. *The OECD Edinburgh Conference on the Scientific and Health Aspects of Genetically Modified Foods, 28 Feb 2000*, Chairman's Report. OECD, Paris.

Persley, G.J. (1990) *Beyond Mendel's Garden: Biotechnology in the Service of World Agriculture*. Biotechnology in Agriculture No. 1, CAB International, Wallingford, UK.

Persley, G.J. (2000) Agricultural biotechnology and the poor: Promethean science. In: Persley, G.J. and Lantin, M.M. (eds) *Agricultural Biotechnology and the Poor: Proceedings of an International Conference, Washington, DC, 21–22 October 1999*. Consultative Group on International Agricultural Research, Washington, DC, pp. 3–21.

Persley, G.J. and Doyle, J.J. (1999) *Biotechnology for Developing Country Agriculture: Problems and Opportunities – Overview*. International Food Policy Research Institute, Washington, DC. (Brief 1 of 10.)

Persley, G.J. and Siedow, J.N. (1999) *Applications of Biotechnology to Crops: Benefits and Risks*. Issue Paper No. 12, Council for Agricultural Science and Technology.

Pinstrup-Andersen, P. and Cohen, M.J. (2000) Modern biotechnology for food and agriculture: risks and opportunities for the poor. In: Persley, G.J. and Lantin, M.M. (eds) *Agricultural Biotechnology and the Poor: Proceedings of an International Conference, Washington, DC, 21–22 October 1999*. Consultative Group on International Agricultural Research, Washington, DC, pp. 159–169.

Platais, G.H. and Persley, G.J. (eds) (2001) *Biodiversity and Biotechnology: Contributions to and Consequences for Agriculture and the Environment*. Environment Department, World Bank, Washington, DC.

Sampaio, M.J.A. (2000) Brazil: biotechnology and agriculture. In: Persley, G.J. and Lantin, M.M. (eds) *Agricultural Biotechnology and the Poor: Proceedings of an

*International Conference, Washington, DC, 21–22 October 1999*. Consultative Group on International Agricultural Research, Washington, DC, pp. 74–78.

Serageldin, I. and Persley, G.J. (2000) *Promethean Science: Agricultural Biotechnology, the Environment and the Poor*. Consultative Group on International Agricultural Research, Washington, DC, 48 pp.

Sharma, M. (2000) India: biotechnology research and development. In: Persley, G.J. and Lantin, M.M. (eds) *Agricultural Biotechnology and the Poor: Proceedings of an International Conference, Washington, DC, 21–22 October 1999*. Consultative Group on International Agricultural Research, Washington, DC, pp. 51–57.

Sittenfeld, A., Espinoza, A.M., Munoz, M. and Zamora, A. (2000) Costa Rica: challenges and opportunities in biotechnology and biodiversity. In: Persley, G.J. and Lantin, M.M. (eds) *Agricultural Biotechnology and the Poor: Proceedings of an International Conference, Washington, DC, 21–22 October 1999*. Consultative Group on International Agricultural Research, Washington, DC, pp. 79–89.

Skerritt, J.H. (2000) Genetically modified plants: developing countries and the public acceptance debate. *AgBiotechNet*, ABN 040.

Swaminathan, M.S. (2000) Genetic engineering and food security: ecological and livelihood issues. In: Persley, G.J. and Lantin, M.M. (eds) *Agricultural Biotechnology and the Poor: Proceedings of an International Conference, Washington, DC, 21–22 October 1999*. Consultative Group on International Agricultural Research, Washington, DC.

Tzotzos, G.T. and Skryabin, K.G. (eds) (2000) *Biotechnology in the Developing World and Countries in Economic Transition*. CAB International, Wallingford, UK.

US National Academy of Sciences (2000) *Transgenic Plants and World Agriculture*. White Paper issued by the National Academy Press, July 2000. Washington, DC, 40 pp.

Wolfenberger, L.L. and Phifer, P.R. (2000) The ecological risks and benefits of genetically engineered plants. *Science* 290; 2088–2093.

World Bank (1997) *1997 World Development Indicators*. World Bank, Washington, DC.

World Bank, (2000) *Food Safety and Developing Countries*. Agricultural Technology Notes No. 26, Rural Development Department, World Bank, Washington, DC, 4 pp.

Zhang, Q. (2000) China: agricultural biotechnology opportunities to meet the challenges of food production. In: Persley, G.J. and Lantin, M.M. (eds) *Agricultural Biotechnology and the Poor: Proceedings of an International Conference, Washington, DC, 21–22 October 1999*. Consultative Group on International Agricultural Research, Washington, DC, pp. 45–50.

# II

# Asia/Pacific

# China

**2**

## Qifa Zhang

| | | | |
|---|---|---|---|
| Area ($km^2$) | 9.597m. | GDP per head | US$3600 |
| Cropland | 10% | | |
| Irrigated cropland | 500,000 $km^2$ | Real GDP growth | |
| Permanent pasture | 43% | (1998 est.) | 7.8% |
| | | Inflation (1998 est.) | 0.8% |
| Population (1999 est.) | 1.246 bn. | | |
| Population per $km^2$ | 129 | Agriculture as % of GDP | 19% |
| Population growth rate | | Value of agricultural | |
| (1999 est.) | 0.7% | exports | 15.5 bn |
| Life expectancy (men) | 69 yrs | Major export commodities: | rice |
| (women) | 72 yrs | | |
| | | Major commodities: | |
| Adult literacy | 81.5% | rice, wheat, potato, sorghum, | |
| | | groundnut, tea, millet, barley, | |
| GDP (1998 est.) | US$4.42 tr. | cotton, oil-seed, pork, fish | |

## Summary

Increasing food production has always been the highest agricultural priority in China, because of the large population. With decreasing arable farmland, food production continues to be a serious problem. The population is expected to increase to 1.6 billion by 2030, and demand for food will increase by at least 60% to keep pace. This population growth and vast urbanization will result in loss of valuable farmland and other natural resources. The only viable approach is to increase the productivity of existing farmland. Statistics show, however, that the production rates of major grain crops have been decreasing in the past decade.

The development of biotechnology has opened up new opportunities to overcome these production problems. China's long tradition in biotechnology research (mainly in tissue culture) will enhance our efforts in advanced biotechnology.

## Introduction

Global commercial production of transgenic crops has increased rapidly in the last few years (James, 1998). There is considerable research and development (R & D) in agricultural biotechnology in China, especially in crop improvement and production.

There is also a huge demand for quality improvement of food products, especially the grain quality of cereal crops. Quality improvement of rice, for example, was largely neglected in breeding programmes in recent years. High yield of cultivars and hybrids is frequently associated with poor quality; most of the widely used cultivars and hybrids have poor cooking and eating qualities and thus are not favoured by producers or consumers.

Another major problem is degradation of the environment. We have seen increasingly frequent natural disasters, such as floods, drought, insect pests and diseases and also expanding areas of soil desertification, salinity and acidity. Extensive applications of chemicals have created a vicious circle in which the excessive use of the chemicals has resulted in a rapid deterioration of the environment and this deterioration has made crop production even more dependent on chemicals.

The greatest challenge is to increase food production and improve product quality in an environmentally sustainable manner.

## Opportunities

In the past 15 years there have been rapid developments in China in scientific infrastructure and also research programmes in biotechnology and molecular biology of various crop plants. Infrastructure developments include the establishment of national key laboratories in the general areas of agricultural biotechnology and crop genetics and breeding, in north, central and south China. These laboratories are well equipped for biotechnology and molecular biology research. In addition, there are open laboratories supported by the Ministry of Agriculture, the Ministry of Education and the Chinese Academy of Sciences. These laboratories have provided good opportunities for biotechnology research.

In the 1990s, regular funding channels were formed at the central government level, which support basic and applied research. This includes the establishment of the National Natural Science Foundation of China and the Chinese Foundation of Agricultural Scientific Research and Education. Major research initiatives and programmes were also established at the state level and by various ministries. The most important programmes for biotechnology R & D are the National Programme on High Technology Development (also known as the 863 Programme) and the National Programme on the Development of Basic Research (also known as 973), both of which included agricultural biotechnology as a major component. Programmes were set up to promote young scientists by awarding special grants from the National Natural Science Foundation, 863 Programme and various ministries. Similar systems, although smaller, were also developed by local governments in many provinces. International funding channels also opened to Chinese scientists during this period, including the Rockefeller Foundation, the McKnight Foundation, the International Foundation for Science and the European Union–China collaboration programmes. The availability of financial support has enhanced research capacity and has promoted the development of young scientists. Some of the programmes have a training component as well.

Rapid advances were made in molecular biology and biotechnology research in China in the 1990s. These include genomic studies in rice and other cereals, development of molecular marker technologies, and identification, mapping and molecular cloning of a large number of agriculturally useful genes. These studies have resulted in powerful tools for varietal improvement (for example, marker-assisted selection), which can be applied to develop new cultivars and hybrid parents.

Transformation technologies have also been established in many laboratories for most of the crop species, including maize, rice and wheat, which are often considered difficult to transform. Transgenic plants can now be routinely produced for rice, maize, wheat, cotton, tomato, potato, soybean, rape-seed and other crops, using *Agrobacterium*, particle bombardment or other methods.

The most up-to-date molecular technologies necessary for varietal development are now in place in China.

Genome mapping and biotechnology research offer powerful tools in crop improvement, including genetic transformation and molecular marker-assisted selection. These techniques offer opportunities to meet the challenges of increased food production.

China's biotechnology programmes now focus on the following priorities (Jiewei, 2000):

- New varieties of high-yielding, high-quality plants and animals resistant to biotic and abiotic stresses, with special emphasis being given to mapping the rice genome.
- New medicines, vaccines and gene therapy technologies.
- Protein engineering technology.

## Disease resistance

More than 20 genes for resistance to various plant diseases have been isolated in recent years (Baker *et al.*, 1997). Analyses of the DNA sequences indicate that the genes share many structural characteristics in common, despite the fact that diseases are caused by a variety of pathogens, such as fungi, bacteria, viruses and nematodes. The genes were isolated from a wide range of plant species, comprising monocotyledonous and dicotyledonous species, including tomato, rice, tobacco and barley. These have provided a rich source of disease-resistance genes for improving resistance by genetic engineering.

Large numbers of genes have been tagged and mapped using molecular markers in many crop species (Zhang and Yu, 1999). Closely linked markers flanking both sides of the genes were identified in many cases. These closely linked markers can be used as the starting-points for isolating the genes, using the map-based cloning approach. These markers can also be used as selection criteria in breeding programmes to monitor the transfer of the genes, which is referred to as marker-assisted selection. New crop lines with improved resistance have been obtained using both approaches.

## Insect resistance

Genes for resistance to various insects have been identified in many crop species and their wild relatives, including gall midge and brown planthopper resistance in rice and pink borer resistance in cotton. Some insect-resistance genes have also been genetically tagged and mapped using molecular markers (Zhang and Yu, 1999). These genes can be directly used in crop breeding programmes using marker-assisted selection.

An important strategy in the development of insect-resistant crop varieties is the utilization of exogenous genes, including genes coding for the endotoxin of *Bacillus thuringiensis* (Bt) and proteinase inhibitors from various sources. Some of the genes have demonstrated strong insecticidal activities under both laboratory and field conditions. Several genes have now been widely used in transformation studies. Many insect-resistant transgenic cotton, maize and rice plants have been produced from these transformation studies, which have now been advanced to the stage of commercial production (James, 1998).

Large-scale utilization of the insect-resistance genes in crop production will not only reduce labour and costs of production; it will also have long-term beneficial effects on the environment. These insect-resistant crops may have a major role to play in sustainable agricultural systems.

## Tolerance to abiotic stresses

Drought and soil salinity and acidity are among the most important threats to agricultural production, causing severe yield losses of all major food crops worldwide. In China, the north-west region is prone to drought, so water supply is a major limitation for crop production; in south and central China, soil acidity is a major limiting factor reducing crop yield; salinity occurs in large areas in the east coastal region.

Drought resistance has been the subject of many studies in several major food crops, including rice, maize and sorghum (Nguyen *et al.*, 1998). Although many quantitative trait loci (QTLs), which explain certain genetic variations in drought tolerance in experimental populations, have been identified by molecular marker mapping, they are unlikely to have a major role to play in improving the drought tolerance of crops.

There have also been QTL studies on the tolerance of rice to acidic soil conditions, especially with respect to aluminium and ferrous iron toxicity (Wu *et al.*, 1999), showing that major gene loci may be involved in increasing the tolerance of rice plants. This may present an opportunity for using genes from rice itself to improve the tolerance of rice varieties to acidic soils.

A more promising line of research is the use of gene coding for citrate synthase, the enzyme for biosynthesis of citric acid (de la Fuente *et al.*, 1997). Transgenic sugar-beet plants with elevated expression of this gene show an enhanced tolerance to aluminium and also increased uptake of phosphate in the acidic soil as a result of excretion of citrate. This indicates that genetic engineering may be able to produce plants that can grow better in acidic soil even with reduced application of phosphate fertilizers. This work may have tremendous implications in crop improvement, especially for crops grown in tropical and subtropical regions.

## Product quality

Biotechnology may have a lot to offer in the improvement of product quality. In rice, for example, the poor cooking and eating qualities of high-yielding cultivars and hybrids represent a major problem for rice production in China. Research has established that the cooking and eating qualities are to a large extent dependent on three traits: amylose content, gelatinization temperature and gel consistency. It was recently shown that all three traits are controlled by the waxy locus located on chromosome 6 (Tan *et al.*, 1999).

The waxy gene was isolated from maize and rice (Shure *et al.*, 1983; Wang *et al.*, 1990). Rice plants transformed with the waxy gene in both sense and antisense configurations showed reduced amylose content, thus demonstrating the usefulness of the transgenic approach in improving cooking and eating qualities. Moreover, the waxy locus has also been clearly defined in the molecular linkage map, and markers residing on the waxy locus and closely linked markers that flank the waxy locus on both sides were identified (Tan *et al.*, 1999). Thus, improvement of the cooking and eating qualities can be achieved using marker-assisted selection.

Another example is the recent success in engineering the entire biochemical pathway for provitamin A biosynthesis (Ye *et al.*, 1999), which significantly enriched vitamin A content in the endosperm of rice grains. This will be a great help to the poor peasant farmers to balance the micronutrients in their diets and hence alleviate malnutrition.

## Increasing yield potential

Several of our major crop species have gone through two great leaps in yield increase in the last several decades: increasing harvest index by reducing the height (making use of the semi-dwarf genes) and utilization of heterosis by producing hybrids. Reduced rates of yield increase have been observed in a number of major food crops in the last 10–15 years (Ministry of Agriculture, 1996). Increasing yield potential has therefore been a common concern in essentially all crop breeding programmes.

Two approaches have been reported in the literature. The first approach is called 'wild QTLs', in which efforts are devoted to bringing QTLs for yield increase from the wild relatives to enhance the yield of cultivars. The argument for such an approach is that only a portion of the genes that ever existed in the wild species was brought to cultivation in the processes of domestication, leaving most of the genes unused. With the help of molecular marker technology, it should therefore be possible to identify genes that can increase the yield of cultivated plants. Xiao *et al.* (1996), for example, reported two QTLs from a wild rice that showed significant effects in increasing the performance of an élite rice hybrid. This has generated considerable interest in identifying genes for agronomic performance from wild relatives that are potentially useful for varietal improvement.

The second approach is to modify certain physiological processes by genetic engineering. Gan and Amasino (1995) reported a system conceived to delay leaf senescence by autoregulated production of cytokinin. The construct was designed by fusing a senescence-specific promoter isolated from *Arabidopsis* with a DNA fragment from *Agrobacterium* encoding isopentenyl transferase (IPT), an enzyme that catalyses the rate-limiting step in cytokinin biosynthesis. The strategy for such a system is that the gene would be turned on at the onset of senescence, leading to the synthesis of cytokinin, and the production of cytokinin would in turn inhibit the process of senescence, thus repressing the expression of this construct itself. Such a system would, therefore, be able to produce cytokinin for delaying senescence, at the same time preventing overproduction of cytokinin, because overproduction of this hormone is detrimental to the plant. Transgenic tobacco plants carrying this construct showed a significant delay in leaf senescence, bringing about a large increase in the number of flowers, number of seeds and biomass, indicating the possibility of increasing plant productivity by delaying leaf senescence. It is interesting, therefore, to determine if this system can provide a general strategy for yield increase in crop improvement.

There are many opportunities for biotechnology to contribute to sustainable food production and to achieve higher yields, better quality and less dependence on chemicals, making crop production more environmentally friendly.

## Constraints

Many constraints still exist that hinder the large-scale research and utilization of transgenic crops.

One of the major constraints relates to intellectual property rights (IPR). China does not yet have effective IPRs for large-scale biotechnology research to develop transgenic crops. Most of the transgenic crop plants that have been developed so far involve complex IPR issues. There is a major shortage of experts in China with knowledge of IPRs and experience in dealing with these issues. China urgently needs help in training people in IPR. Scientists and breeders do not fully understand IPRs, which are often not recognized and honoured. Public education is therefore urgently needed.

Another major constraint is the lack of extension mechanisms that take the products of biotechnology research to the farmers. China had a network system to dispense agricultural technologies, seeds and other related materials, but with the development of a market economy, the old distribution systems are gradually losing their effectiveness and are now evolving into profit-driven seed companies undergoing the processes of privatization. Although this may be a good movement in itself, it may take several years for the system to become effective, because the funding situation does not appear to be promising. Governmental support goes mainly to the research component and there is not enough funding to support initiatives and start-ups of seed companies.

There are also a number of scientific and technical constraints to the application of technology in crop improvement. One of the constraints is the lack of understanding of the mechanisms governing the traits that are important in crop improvement. Drought causes severe yield loss worldwide and it will continue to be among the most damaging stresses in crop production. Tolerance of the crop to drought as a trait, however, has not been well defined and it is still not clear what aspects of plant morphology or physiology are the most important for drought tolerance. Research is still needed to define a clear target for improving drought tolerance.

There is also a need for more germ-plasm. Germ-plasm has not been found for a number of important traits, such as resistance to fungal diseases and resistance to a number of pests in crop species (for example, sheath blight of rice, scab disease of wheat and yellow wilt of cotton). These have become devastating diseases worldwide, as have borer insects of a number of crops, which cause heavy damage. International collaboration, coordinated by the international agricultural research centres, may have a crucial role to play in germplasm identification, exchange and utilization.

### Field testing of transgenic crops

Transgenic research has been conducted on 47 plant species in China, using 103 genes. A national committee for the regulation of biosafety of genetically improved agricultural organisms was established in 1996 to promote biotechnology in a healthy environment. This committee accepts applications twice a year for biosafety evaluation of genetically improved agricultural organisms, such as crop plants, farm animals and microorganisms.

By mid-1998, the committee had received 86 applications, of which 75 were for field testing of transgenic crops. Permission for 53 of the applications was granted for commercial production, environmental release or small-scale field testing (Chinese Society of Agricultural Biotechnology, 1998a, b). The crops used for transgenic research were rice, wheat, maize, cotton, tomato, pepper, potato, cucumber, papaya and tobacco. Traits targeted for improvement included disease resistance, pest resistance, herbicide resistance and quality improvement. In a few cases, transgenic crops have been grown for large-scale commercial production. We expect that the area planted with transgenic crops will increase rapidly in the next few years.

## Conclusions

Recent developments in genome mapping and genetic engineering have provided a knowledge base, identified germ-plasm resources, provided useful genes and offered effective tools for crop improvement. Integration of the knowledge, the tools and the genetic resources into breeding programmes will greatly increase the efficiency of varietal development.

It is expected that molecular marker-assisted selection will have a major role to play in the future genetic improvement of many crops. This is not only because the technique itself has provided a highly efficient tool for speedy and precise selection, but also because it possesses several distinct advantages. First, it does not require the isolation of the targeted gene, which often takes many years and massive resources to accomplish. Secondly, most of the gene constructs such as those commonly used in many transformation studies are now covered by IPRs and therefore are not freely available for varietal development. Thirdly, the progeny developed by marker-assisted selection in general does not suffer from adverse effects such as over- or underexpression and transgene silencing, which are now frequently reported with transgenic plants. The performance of the progeny resulting from marker-assisted selection is therefore much more predictable than that from transformation. The large number of genes that have been precisely tagged and mapped will provide a rich source for marker-assisted breeding.

The most common practice for obtaining new genes is map-based cloning. Molecular markers that are closely linked to genes of interest can serve as the starting-point for cloning the genes, following the map-based cloning approach. It can be expected that the process of gene isolation using this approach will be greatly accelerated with advances in the international effort in DNA sequencing. It is highly likely that all the genes that are accurately mapped with closely linked markers can be quickly isolated with the availability of the sequence information.

The recent development in DNA-chip technologies may also provide a powerful tool for large-scale isolation of new genes in the near future (Lemieux *et al.*, 1998). It can be expected that large numbers of genes will become available for crop improvement in the next decade, which in turn will promote transgenic research.

Biotechnology will soon play a major role in crop improvement in China. The area planted to cultivars developed using biotechnology will increase steadily in the years to come. Biotechnology will contribute significantly to food production and food security in the new century.

## References


Baker, B., Zambryski, P., Staskawicz, B. and Dinesh-Kumar, S.P. (1997) Signaling in plant–microbe interactions. *Science* 276, 726–733.

Chinese Society of Agricultural Biotechnology (1998a) Results of the biosafety evaluation of agricultural organisms. *Agricultural Biotechnology Newsletter* 2, 5–8.

Chinese Society of Agricultural Biotechnology (1998b) Results of the biosafety evaluation of agricultural organisms. *Agricultural Biotechnology Newsletter* 3, 7–8.

de la Fuente, J.M., Ramirez-Rodriguez, V., Cabrera-Ponce, J.L. and Herrera-Estrella, L. (1997) Aluminium tolerance in transgenic plants by alteration of citrate synthesis. *Science* 276, 1566–1568.

Gan, S. and Amasino, R.M. (1995) Inhibition of leaf senescence by autoregulated production of cytokinin. *Science* 270, 1986–1988.

James, C. (1998) *Global Review of Commercialized Transgenic Crops: 1998*. ISAAA Briefs No. 8, ISAAA, Ithaca, New York.

Jiewei, W. (2000) China. In: Tzotzos, G.T. and Skryabin, K.G. (eds) *Biotechnology in the Developing World and Countries in Economic Transition.* CAB International, Wallingford, UK, pp. 60–63.

Krattiger, A.F. (1997) *Insect Resistance in Crops: a Case Study of* Bacillus thuringiensis (Bt) *and its Transfer to Developing Countries.* ISAAA Briefs No. 2, ISAAA, Ithaca, New York.

Lemieux, B., Aharoni, A. and Schena, M. (1998) Overview of DNA chip technology. *Molecular Breeding* 4, 277–289.

Ministry of Agriculture of PRC (1996) *Report of Agriculture Development in China.* China Agricultural Press, Beijing.

Nguyen, H.T., Babu, R.C. and Blum, A. (1998) Breeding for drought resistance in rice: physiology and molecular genetics considerations. *Crop Science* 37, 1426–1434.

Shure, M., Wessler, S. and Federoff, N. (1983) Molecular identification and isolation of the *waxy* locus in maize. *Cell* 35, 225–233.

Tan, Y.F., Li, J.X., Yu, S.B., Xing, Y.Z., Xu, C.G. and Zhang, Q. (1999) The three important traits for cooking and eating quality of rice grains are controlled by a single locus in an élite rice hybrid, Shanyou 63. *Theoretical and Applied Genetics* 99, 642–648.

Wang, Z.Y., Wu, Z.L., Xing, Y.Y., Zheng, F.Q., Guo, X.L., Zhang, W.G. and Hong, M.M. (1990) Nucleotide sequence of rice *waxy* gene. *Nucleic Acids Research* 18, 5898.

Wu, P., Luo, A., Zhu, J., Yang, J., Huang, N. and Senadhira, D. (1997) Molecular markers linked to genes underlying seedling tolerance for ferrous iron toxicity. In: Ando, T. *et al.* (eds) *Plant Nutrition for Sustainable Food Production and Environment*, Kluwer Academic Publishers, Dordrecht, Netherlands, pp. 789–792.

Xiao, J., Grandillo, S., Ahn, S.N., McCouch, S.R., Tanksley, S.D., Li, J. and Yuan, L. (1996) Genes from wild rice improve yield. *Nature* 384, 223–224.

Ye, X., Al-babili, S., Klöti, A., Zhang, J., Lucca, P., Beyer, P. and Portrykus, I. (2000) Engineering the provitamin A (β-carotene) biosynthetic pathway into (carotenoid-free) rice endosperm. *Science* 287, 303–305.

Zhang, Q. and Yu, S. (1999) Molecular marker-based gene tagging and its impact on rice improvement. In: Nanda, J.S. (ed.) *Rice Breeding and Genetics – Research Priorities and Challenges*. Science Publishers, Enfield, New Hampshire, pp. 241–270.

# India

## 3

Manju Sharma

| | |
|---|---|
| Area (km²) | 3.287m. |
| Cropland | 46% |
| Irrigated cropland | 40.6% |
| Permanent pasture | 4% |
| Population (2001 est.) | 1.012 bn |
| Ann. pop. growth (2001 est.) | 1.79% |
| Life expectancy (men and women) | 62 yrs |
| Adult literacy (1991) | 52.2% |
| GDP (1999 est.) | US$385 bn |
| GDP per head | US$400 |
| Growth in real GDP (1999) | 6.2% |
| Agriculture as % of GDP | 26.6% |
| Value of agricultural and marine exports (1998/99) | US$6 bn |
| Agriculture and marine products as % of total exports | 17.8% |
| Major export commodities: rice, tea, jute, cashew nut, coffee, spices, marine products | |

## Summary

In the 1980s a number of scientific agencies in India were supporting research in modern biology: the Council of Scientific and Industrial Research (CSIR), the Indian Council of Agricultural Research (ICAR), the Indian Council of Medical Research (ICMR), the Department of Science and Technology (DST) and the University Grants Commission, among others. The establishment by the Indian government of the National Biotechnology Board in 1982 gave biotechnology an important boost. Training, creation of infrastructure facilities and supporting research and development (R & D) in carefully identified areas were given highest priority. The success and impact of the National Biotechnology Board prompted the government to establish a separate Department of Biotechnology (DBT) in February 1986, resulting in major accomplishments in basic and applied research in agriculture, health, the environment, aquaculture and marine biotechnology and also in areas of training, industry, biosafety and bioethics.

## Introduction

Basic research is essential in all aspects of modern biology, including development of the tools to identify, isolate and manipulate individual genes that govern specific characters in plants, animals and microorganisms. Recombinant DNA technology is the basis for these new developments. The creativity of scientists and basic curiosity-driven research will be the keys to future success. Areas of biosystematics using molecular approaches, mathematical modelling and genetics, including genome sequencing for humans, animals and plants, continue to have priority in the country. The impact of genome sequencing is increasingly evident in many fields. New avenues are opening in the biosciences with the power of high-throughput sequencing and rapidly accumulating sequencing data.

In plant genomics, the sequencing of *Arabidopsis* has now been completed and published (Arabidopsis Genome Initiative, 2000) and the rice genome will soon be completed and catalogued. There have been major achievements in basic bioscience in the last decade in India and we have developed expertise in practically all areas of molecular biology (see also Sharma and Swarup, 2000). The institutions under the CSIR, ICMR, ICAR, DST and DBT have established a large number of facilities where most advanced research work in the biosciences is carried out. Considerable success has been achieved in the identification of new genes, development of new drug delivery systems, diagnostics, recombinant vaccines, computational biology and many other related areas. Breakthroughs include studies on the three-dimensional structure of a novel amino acid and a long-chain protein of the mosquito (University of Pune) and a demonstration of the potential of the reconstituted Sendai viral envelopes containing only the F protein of the virus as an efficient and site-specific vehicle for the delivery of reporter genes into hepatocytes (Delhi University).

## Agriculture and Allied Areas

The post-green-revolution era is almost merging with the gene revolution to improve crop productivity and quality. The Indian research agenda will for some years be dominated by the exploitation of heterosis vigour and the development of new hybrids, including apomixis, genes for abiotic and biotic resistance, developing planting material with desirable traits and the genetic enhancement of important crops. Integrated nutrient management and development of new biofertilizers and biopesticides would be important contributions to sustainable agriculture, soil fertility and a clean environment. Stress biology, marker-assisted breeding programmes and studying the important genes will continue as priorities.

We have achieved the cloning and sequencing of at least six genes, developed regeneration protocols for citrus, coffee and mangrove species and new types of biofertilizer and biopesticide formulations, including mycorrhizal fertilizers. Research to develop transgenic plants for brassicas, mung bean, cotton and potato is well advanced. Industries have shown a keen interest in the applications of biotechnology, including field trials, and have set up production facilities for biopesticides, biofertilizers and tissue-culture plants. The success of the tissue-culture pilot plants in the country, at Tata Energy Research Institute in New Delhi and the National Chemical Laboratory in Pune, are now functioning as Micropropagation Technology Parks. This has given a new direction to the plant tissue-culture industry. The micropropagation parks serve as a platform for the effective transfer of technology to entrepreneurs, including training and the demonstration of technology for mass multiplication of horticulture crops and trees. Considerable research progress has been made with cardamom and vanilla, both important crops. Cardamom yields have increased 40% using tissue-cultured plants.

While the green revolution gave us self-reliance in food, the livestock population provided a 'white revolution', making India the largest milk-producing country in the world. With 80% of the milk in India coming from small and marginal farms, progress in milk production has had a major social impact. A diverse infrastructure has been established to help farmers in the application of embryo-transfer technology. The world's first *in vitro* fertilization (IVF) buffalo calf was born through embryo-transfer technology at the National Dairy Research Institute, Karnal. Multiple ovulation and embryo transfer, *in vitro* embryo production, embryo sexing, vaccines and diagnostic kits for animal health have also been developed. Waste recycling technologies that are cost-effective and environmentally safe are being generated. The animal science area is also opening up many avenues for employment generation.

With an extensive coastline, India has great potential for marine resource development and aquaculture. Scientific aquaculture offers real opportunities to achieve an annual target production of 10 million metric tonnes of fish. Aquaculture products are among the fastest-moving commodities in the world, so we have to continuously improve seed production, feed, health products, cryopreservation methods, genetic studies and related environmental factors. Aquaculture will help substantially in the diversification of the bread basket and in combating nutritional deficiency.

## Food security

Biotechnology offers significant promise to provide healthier and more nutritious food. Millions of people are malnourished and vitamin A deficiency affects 40 million children. There are also serious deficiencies of iodine, iron and other important nutrients. Every year over 6 million children under the age of 5 die worldwide. About 2.7 million of these children die in India. More than half of these deaths result from inadequate nutrition.

With the advent of gene transfer technology and its use in food crops, we hope to achieve not only higher productivity, but better quality, including nutrition and storage properties. We also hope to ensure the adaptation of plants to specific environmental conditions, to increase plant tolerance to stress conditions, to increase pest and disease resistance and to achieve higher prices in the market-place. Genetically modified foods will have to be developed under adequate regulatory processes, with full public understanding. We should ensure proper labelling of the genetically modified foods, so that consumers will have a choice.

It is scientifically established that an environmentally benign way of ensuring food security is through the bioengineering of crops. For the 4.6 billion people in developing countries, presuming that 1 billion do not get enough to eat and live in poverty, is there any other strategy or alternative? Biotechnology will provide the new tools for breeders to enhance crop productivity, as well as its nutritional quality.

## Plant biotechnology

India has two hot spots of biodiversity and more than 47,000 species of plants which constitute 8% of the total diversity on the earth. This bioresource constitutes the mainstay of the economy of the poor people, so special emphasis is required for plant biotechnology research. Isolation of genes for abundant proteins, combining molecular genetics and chromosome maps, and a much better understanding of the evolutionary relationship of the members of the plant kingdom have demonstrated the potential of plants as major sources of food, feed, fibre, medicine and industrial raw materials. Molecular fingerprinting and areas of genomics and proteomics will accelerate our efforts in developing suitable transgenic crops. By identifying appropriate determinants of male sterility, we can extend the benefit of hybrid seeds to more crops. We must help the farmer by ensuring hybrid vigour generation after generation. Additional research on apomixis would open up such possibilities.

A National Plant Genome Research Centre has been established at Jawaharlal Nehru University. A number of centres for plant molecular biology established in different parts of the country were initially responsible for training in crop biotechnology. We are emphasizing plant molecular and biotechnology research because of the potential to produce more proteins, vitamins, pharmaceuticals, colouring material, bioreactors, edible vaccines, therapeutic antibodies and drugs.

## Environment

The global scientific community is concerned about environmental protection and conservation and policies of sustainable development and environmental protection. The Stockholm Conference in 1972 and the Conference in Rio de Janeiro in 1992 focused world attention on areas of pollution, biodiversity conservation and sustainable development. Plants and microbes are becoming important in pollution control. World Bank estimates show that pollution in India is costing almost US$80 billion in terms of sickness and death. New developments, such as bioindicators, phytoremediation methods, bioleaching, biosensors and the identification and isolation of microbial consortia for regeneration of degraded and waste lands, are becoming priority research areas. Although significant work has been done in India, developing a more plant-orientated approach to pollution control is extremely important. Cleaning up the large river systems and ensuring the removal of pesticide residue in large urban slums are priorities in which a biotechnological approach would be environmentally safe.

Phytoremediation to remove the high levels of explosives found in the soil has become a reality. Although it was known that some microbes can denitrify nitrate explosives in the laboratory, they could not thrive on site. French *et al.* (1999) have transferred this degradative ability from the microbe to tobacco plants and these have produced a microbial enzyme capable of removing the nitrates.

## Biodiversity

The global biosphere can survive only if resource utilization is about 1% and not 10%. The global environment is regulated by climate changes and biosphere dynamics. Knowledge about biodiversity is being used by scientists throughout the world and we have many gene banks, botanical gardens and herbaria for conservation purposes. There are also molecular approaches, including DNA fingerprinting, for plant characterization and conservation. All species and ecosystems have become exceedingly important, not only for understanding the global environment but also from the viewpoint of the enormous commercial significance of biodiversity.

Biotechnology is becoming a major tool in conservation biology. Twelve per cent of vascular plants are threatened with extinction and 5025 animal species are threatened worldwide, including 563 Indian species. India has about 2000 species of vascular plants under threat.

Biodiversity is under serious threat and understanding the scale of this destruction and extinction is essential. Nations must address questions such as who owns the biodiversity, who should benefit from it and what the role of society and the individual is in protecting biodiversity.

More research is needed on forests, marine resources, bioremediation methods, restoration ecology and large-scale tree plantations. The latter have reached 180 million ha and may increase substantially in the next decade. Marine resources provide many goods and benefits, including bioactive materials, drugs and food items, which must be characterized and conserved.

## Medical Biotechnology

A major responsibility of biotechnologists in the 21st century will be to develop low-cost, affordable, efficient and easily accessed health-care systems. Advances in molecular biology, immunology, reproductive medicine, genetics and genetic engineering have revolutionized our understanding of health and diseases and we may be entering an era of predictive medicine. Genetic engineering promises to treat a number of monogenetic disorders and to unravel the mystery of polygenetic disorders with the help of research on genetically improved animals.

Every year millions of people die of infectious diseases. The main killers according to the World Health Organization (WHO) are acute respiratory infection, diarrhoeal diseases, tuberculosis, malaria, hepatitis and HIV-AIDS.

There are new vaccines being developed for many diseases, and diagnostic kits for HIV, pregnancy detection and hepatitis are being developed as well. These technologies have been transferred to industry.

Guidelines for conducting clinical trials for recombinant products have been developed, which have now been accepted by the Health Ministry and circulated widely to industry. Promising leads now exist to develop vaccines for rabies, *Mycobacterium* tuberculosis, cholera, JEV and other diseases. A recombinant hepatitis B vaccine, LEPROVAC, is already on the market. There is a Jai Vigyan technology mission on the development of vaccines and diagnostics. A National Brain Research Centre is being established to improve our knowledge of the human brain and brain diseases.

The discovery of new drugs and development of the drug delivery system are increasingly important. Bioprospecting for important molecules and genes for new drugs has already begun as a multi-institutional effort. A recombinant vaccine for Bacillus Calmette-Guérin (BCG) and hepatitis is being developed. The age-old system of Ayurveda practised in India needs to be popularized and made an integral part of health care.

## Industrial Biotechnology

Advances in biotechnology can be converted into products, processes and technologies by developing an interdisciplinary team. The pharmaceutical sector has had a major impact in this field, as rare therapeutic molecules in the pure form become available. Diagnostics have mushroomed and over 600 biotechnology-based diagnostics are now available in clinical practice, with a value of about US$20 billion. Polymerase chain reaction-based diagnostics are the most common. The Indian effort in the diagnostic area has been commendable and it is expected that sales will be about US$450 million in the early 2000s.

The consumption of biotechnology products was expected to increase from US$600 million to about US$1.2 billion by 2000. Industrial enzymes have emerged as a major vehicle for improving product quality. In India a number of groups are gearing up to produce industrial enzymes such as $\alpha$-amylose, proteases and lipases, with an anticipated threefold increase by the end of 2000. India is now producing 13 antibiotics by fermentation. Capacity exists to produce important vaccines, such as DPT, BCG, JEV, cholera and typhoid. Cell-culture vaccines, such as MMR, rabies and hepatitis B, have also been introduced.

## Bioinformatics

The coming together of biotechnology and informatics is paying rich dividends. Genome projects, drug design and molecular taxonomy are all becoming increasingly dependent on information technology. Information on nucleotides and protein sequences is accumulating rapidly. The number of genes characterized from a variety of organisms and the number of evolved protein structures are doubling every 2 years. The DBT has established a countrywide Bioinformatics Network, with ten Distributed Information Centres (DICs) and 35 sub-DICs. A Jai Vigyan Mission on the establishment of genomic databases has been started, with a number of graphic facilities created throughout the country. This system has helped scientists involved in biotechnology research.

## Bioethics and Biosafety

The bioethics committee of the United Nations Educational, Scientific and Cultural Organization (UNESCO), established in 1993, has produced a number of guidelines for ethical issues associated with the use of modern biotechnology.

Biosafety guidelines for genetically modified organisms need to be strictly followed to prevent any possible harm to the environment. A three-tier mechanism of Institutional Biosafety Committees has been instituted in India: the Review Committee on Genetic Manipulation, the Genetic Engineering Approval Committee and the state-level coordination committees. It is important to give a clear explanation of the new biotechnologies to the public to allay fears. New models of cooperation and partnership have to be established to ensure close linkages among research scientists, extension workers, industry, the farming community and consumers.

Gene transformation is done worldwide with four broad objectives: (i) to develop products with new characteristics; (ii) to develop pest and disease resistance; (iii) to improve nutritional value; and (iv) to modify fruit ripening to obtain longer shelf-life. Thus the aims and objectives are laudable and the tools are available. It does, however, call for a cautious approach following appropriate biosafety guidelines.

About 25,000 field trials of genetically modified crops have been conducted worldwide. The anticipated benefits are better planting material, savings on inputs and genes of different varieties that can be introduced in the gene pool of crop species for their improvement. The potential risks include weediness, transgene flow to non-target plants and the possibility of new viruses developing with wider host range and their effects on unprotected species. For crops such as maize and cotton with single-gene introductions, there is very little problem expected. When multiple genes are involved, scientists have to be more cautious.

The time has arrived for a serious look at the ethical and biosafety aspects of biotechnology. Researchers, policy-makers, non-governmental organizations (NGOs), progressive farmers, industrialists, representatives of the government and all concerned need to come together and share a platform to address the following issues:

- Environmental safety.
- Food and nutrition security.
- Social and economic benefits.
- Ethical and moral issues.
- Regulatory issues.

## Training

There are more than 50 approved Master of Science (MS), postdoctoral and Doctor of Medicine (MD) training programmes in biotechnology in different institutions and universities, covering most Indian states. Short-term training programmes, technician training courses, fellowships for students to go abroad, training in Indian institutions, popular lecture series, awards and incentives form an integral part of the training activities in India. A special feature of the programme has been that, since 1996, many students, after completion of their training course, join industries or work in biotechnology-based programmes in institutions and laboratories. National Bioscience Career Development Awards have been instituted. Special awards for women scientists and scholarships for the best students in biology help promote biotechnology in India and give recognition and reward to the scientists.

## Special Programmes

Biotechnology-based activities to benefit the poor and programmes for women have been launched. A unique feature is the establishment of a Biotechnology Golden Jubilee Park for Women, which will encourage a number of women entrepreneurs to take up biotechnology enterprises.

States are taking a keen interest in developing biotechnology-based activities. The states of Uttar Pradesh, Arunachal Pradesh, Madhya Pradesh, Kerala, West Bengal, Jammu and Kashmir, Haryana, Mizoram, Punjab, Gujarat, Meghalaya, Sikkim and Bihar have already started large-scale demonstration activities and training programmes.

## Investment

The Indian government has made substantial investments in biotechnology research. Bringing Indian biotechnology products to market will require the involvement of large and small entrepreneurs and business houses. This will require substantial investments from Indian and overseas investors. The worldwide trend is towards companies becoming major players in the development of biotechnology products and in supporting product-related biotechnology research.

## Conclusions

In the years ahead, biotechnology R & D should produce a large number of new genetically improved plant varieties in India, including cotton, rice, brassicas, pigeon pea, mung bean and whcat. Tissuc-culture regeneration protocols for important species, such as mango, saffron, citrus and neem will be available for large-scale planting. Micropropagation technology will provide high-quality planting materials for farmers. Environment-friendly biocontrol agents and biofertilizer packages will hopefully be made available to farmers in such a way that they can produce these in their own fields. The country should be in a position to fully utilize, on a sustainable basis, medicinal and aromatic plants. The development through molecular biology of new diagnostic kits and vaccines for major diseases would make the health-care system more efficient and cheaper. Genetic-counselling clinics, molecular probes and fingerprinting techniques should all be used to solve the genetic disorders in the population. The establishment of *ex situ* gene banks to conserve valuable germ-plasm and diversity and a large number of repositories and referral centres for animals, plants and microorganisms should be possible. Detailed genetic read-outs of individuals could be available. Information technology and biotechnology together should become a major economic force. It is expected that plants as bioreactors would be able to produce large numbers of proteins of therapeutic value and many other important items. *In vitro* mass propagation can be carried out on any desired species with non-random programming. Certainly the 21st century could witness a major increase in new bioproducts generated through modern biology.

To achieve our goal of self-reliance in this field, India will require a strong educational and scientific base, a clear public understanding of new biotechnologies and the involvement of society in many of these biological ventures. India has a large research and educational infrastructure, comprising 29 agriculture universities, 204 central and state universities and more than 500 national laboratories and research institutions. It should therefore be possible to develop capabilities and programmes so that these institutions act as regional hubs for the farming community, where they can get direct feedback about new technological interventions. It will be equally important to establish strong partnerships and linkages with industry, from the time a research lead has emerged until the packaging of the technology and commercialization are achieved.

## References

Arabidopsis Genome Initiative (2000) Analysis of the genome sequence of the flowering plant *Arabidopsis thaliana*. *Nature* 408 (14 December), 796–815.

French, C.E., Rosser, S.J., Davies, G.J., Nicklin, S. and Bruce, N.C. (1999) Biodegradation of explosives by transgenic plants expressing pentaerythritol tetranitrate reductase. *Nature Biotechnology* 17 (5), 491–494.

Sharma, M. and Swarup, R. (2000) India. In: Tzotzos, G.T. and Skryabin, K.G. (eds) *Biotechnology in the Developing World and Countries in Economic Transition.* CAB International, Wallingford, UK, pp. 85–88.

# Indonesia

**4**

**P.J. Dart, I.H. Slamet-Loedin and E. Sukara**

| | |
|---|---|
| Area ($km^2$) | 1.919m. |
| Cropland | 10% |
| Irrigated cropland | 45,970 $km^2$ |
| Permanent pasture | 7% |
| Population (1999 est.) | 216.1m. |
| Population per $km^2$ | 88 |
| Ann. pop. growth rate (1998) | 1.46% |
| Life expectancy (men) | 60.6 yrs |
| (women) | 65.3 yrs |
| Adult literacy | 83% |
| GDP (1998 est.) | US$602 bn |
| GDP per head | US$2830 |
| Growth in real GDP (1998) | 13% |
| Inflation (1998 est.) | 77% |
| Agriculture as % of GDP | 18.8% |
| Value of agricultural exports | US$1.8 bn. |
| Agricultural products as % of total exports | > 6.2% |

Major export commodities: rubber, palm oil, shrimp

Major subsistence commodities: rice, cassava, banana

## Summary

Indonesia has placed a high priority on the development of biotechnology. The State Minister of Research and Technology has designated four national centres of excellence to coordinate R & D in agricultural, medical and industrial biotechnology.

Applications of biotechnology in agriculture are the responsibility primarily of AARD, carried out at eight research institutes. AARD (1990) published a major study detailing its proposed Programme of Research in Agricultural Biotechnology. *A Blueprint for the Development of Biotechnology in Indonesia* (Office of the State Minister for Research and Technology, 1985), laid the framework for biotechnology development. A National Committee for Biotechnology was established to assist the Minister in coordinating R & D and in developing guidelines for government policy in the promotion of biotechnology. A major personnel training programme within Indonesia and abroad is in progress.

## Introduction

Having achieved self-sufficiency in rice production, Indonesia initiated changes in research and development (R & D) policy during the 1984–1989 five-year plan (Repelita IV) and these continued during 1989–1994 (Repelita V). There was increased emphasis on the non-rice (*palawija*) crops (maize, soybean, cassava and other tuber crops, groundnut), horticultural and estate crops, livestock and fisheries, and an increasing focus on farming systems for different agroecological zones (Research Highlights and Research Reviews, 1991). Considerable increases in annual production were targeted for rice (2.4%), maize (3.5%), soybean and groundnuts (10%), estate and industrial crops (5% for rubber and coconut to 14% for oil-palm and 22% for cocoa). There is an increased emphasis on self-sufficiency, diversification, nutrition, increased employment and incomes, especially through value-added agribusiness enterprises, with the nucleus estate as one model, improved use of natural resources and environmental protection. The economic structure was changed to give increased emphasis to equal development between agriculture and industry and the development of science and technology to enhance the ability to use advanced technology for long-term development.

A Workshop on Biotechnology and Agroindustry sponsored by the Office of the Minister of State for Research and Tcchnology and the US National Academy of Sciences, held in Jakarta in 1983, led to the concept of a centre for R & D for biotechnology at Cibinong as part of the National Programme of Development of Biotechnology. The Ministry of State for Research and Technology has shown interest in the development of biotechnology in Indonesia, particularly agricultural biotechnology, emphasizing the potential role in increasing and stabilizing food production in the less favourable environments of the outer islands and increasing yields through control of diseases, as well as the income-generating and import-substitution role for biotechnology to generate Indonesian products in agroindustry businesses.

Under the National Programme, three Inter-University Centres (IUCs) on Biotechnology were established in 1985: Bogor Agricultural University (IPB) – agricultural biotechnology; Bandung Institute of Technology (ITB) – industrial biotechnology; and the University of Gadjah Mada (UGM) – medical biotechnology. These IUCs help train the skilled scientists and technicians required nationally to undertake strategic and basic research. Government agencies, industry and universities have organized several workshops on biotechnology, an indication of the growing involvement in biotechnology in Indonesia.

Further, the Agency for the Assessment and Application of Technology (BPPT) was established under the direction of the Minister of State for Research and Technology with headquarters at Serpong, Jakarta, with specific interest in industrial biotechnology applications, such as the processes for production of chemicals (ethanol, enzymes, amino acids), pharmaceuticals (antibiotics, vitamins), plant seedlings (asparagus, rattan, *Solanum*) and single-cell protein feeds.

A National Committee for the Development of Biotechnology (ten members) was established in 1985 to monitor and coordinate the development of biotechnology in Indonesia, including the formulation of regulations on the importation, research, release into the environment and testing of genetically modified organisms and setting R & D priorities and funding.

The Indonesian Institute of Sciences (LIPI) has a major centre for biotechnology – the Research and Development Centre for Biotechnology (RDCBt), Cibinong Bogor – as well as further activity at the Research and Development Centre for Applied Chemistry (PPPKT/RDCAC), Bandung.

The ASEAN Subcommittee on Biotechnology was set up by the ASEAN Committee on Science and Technology (COST) in 1989 to enhance capabilities for R & D in ASEAN countries. A coordinated research programme sponsored by the Australian International Development Assistance Bureau (AIDAB, now AusAID) was established in 1989 with programmes in six countries, including Indonesia (the RDCBt-LIPI, Cibinong Bogor; IUC-ITB and RDCAC-LIPI, Bandung; PT Kimia Farma).

Agriculture plays a substantial role in the Indonesian economy, involving more than 55% of the population, 19% of the gross domestic product (GDP) and more than 60% of the value of non-oil exports. Over the last two decades, annual agricultural output has grown by 4%. Rice production accounts for more than 40% of agricultural output, land use and employment. Production increased from 12 million t in 1969 to 44 million t in 1991, decreasing to 39.9 million t in 1999 (Indonesian statistics). A similarly dramatic increase occurred in livestock production, including fish and eggs, from 2.2 million t in 1977 to 4.3 million t in 1987, a 52% increase. However, the projected increases in population have given increased importance to the role of research in sustaining self-sufficiency in food production.

The Agency for Agricultural Research and Development (AARD) recognizes the role biotechnology must play in research, and several of the central research institutes have increasing activity in this area. The plantation crops and sugar institutes have direct funding from grower and private-company bodies and activities closely linked to industrial production.

The Biotechnology Laboratory of the Central Research Institute for Food Crops (CRIFC), AARD, Bogor, was opened in 1990, with buildings and equipment supported by the Japan International Cooperation Agency (JICA). This laboratory was designated by the Minister of State for Research and Technology as one of the three national centres of excellence, along with industrial biotechnology at BPPT, Serpong and medical biotechnology at the University of Indonesia, Jakarta. These three centres are to take a lead role in the national development of biotechnology, in both conducting and coordinating R & D. In 1993, the Minister of State for Research and Technology designated RDCBt, LIPI, as the second centre of excellence for agricultural biotechnology.

## Biotechnology Policy

In the early 1980s, biotechnology started to receive more attention from Indonesian scientists. Programmes and activities to exploit the potentials of biotechnology were initiated in research agencies/institutes and universities. In 1985 the development of biotechnology was declared a national priority and the Office of the Minister of State for Research and Technology published a government paper entitled *Pattern of Development of Biotechnology in Indonesia*.

This was done for two reasons: (i) biotechnology can contribute substantially to national development; and (ii) biotechnology will give better, cheaper and quicker solutions to many of the development constraints, maybe even offering the only viable solution.

It notes that 'the development of bio-industries and research in biotechnology should be supported by skilled personnel and adequate facilities that will be able to deliver innovations in the future'.

With the above objectives, two targets were identified:

**1.** To have developing industries based upon the application of biotechnology to produce goods and services for human welfare and the growth of innovation.
**2.** To have R & D of biotechnology, in all its aspects, which will support a sustainable bioindustry.

Four phases of biotechnology development were proposed:

**1.** The transfer of technology, where existing biotechnology skills and processes will be imported, with the aim of producing high value-added goods and services. This stage simultaneously provides opportunities for Indonesia to understand the design and techniques of biotechnology.
**2.** The technological integration of biotechnology, which will assist in the formation of new designs for goods and services produced by biotechnology in Indonesia.

**3.** The technological development of biotechnology, during which new Indonesian technology is developed in order to give the developing bioindustry a comparative advantage.
**4.** Local basic research, which would be able to support the development of biotechnology for the ongoing development of bioindustries within Indonesia.

Biotechnology was to be developed to enhance the national capability in the development of health care, agriculture and industry and increase the economic capability of the country. Fulfilling these objectives would involve the following steps:

- Mobilizing national capacity to implement objectives of the *Pattern of Development of Biotechnology in Indonesia.*
- Encouraging and guiding the utilization of biotechnology processes adopted by bioindustries elsewhere, producing high value-added products or services.
- Encouraging and mobilizing biotechnology research and developing a stable research network in Indonesia that would gradually support the development of a bioindustry, with self-sufficiency in innovation for practical commercial biotechnology around 2000.

Guiding the development of a skilled biotechnology workforce, including labour, university graduates and postgraduates, for bioindustries and research institutions is a priority task in Indonesia. Such development should occur with Indonesian nationals both inside and outside the country. This trained workforce will also be crucial for the effectiveness of technology transfer. The government recognizes that the universities' capacities in microbiology, biochemistry, genetics and biology will need strengthening (see also Moeljopawiro, 2000).

A major shift in policy has occurred in the relationships between private industry and government research institutions. Industry is encouraged to develop collaborative projects on a cost-sharing or turnkey basis, i.e. hiring laboratory facilities and researchers, with the government institute having the opportunity to keep the revenue earned to support the institute's overall operating budget. Conversely, government institutions, such as LIPI, are encouraged to compete for contract research from private companies.

To accelerate the implementation of the programme, the government of Indonesia also revitalized the National Research Council. The Council directs biotechnology priorities for each fiscal year and invites scientists from the universities and research institutes (both public and private) to submit research proposals. The Council sets up a panel of experts to evaluate the proposals and make recommendations to the Council. The Council recommends proposals for funding to the National Planning Board and the Ministry of Finance.

Since 1994 the Indonesian government has been providing competitive grants for biotechnology research, resulting in a significant increase in high-quality research activities. The major groups doing biotechnology research are universities and R & D centres of departmental and non-departmental bodies. Various private companies are also conducting biotechnology research in Indonesia.

The National Committee on Biotechnology set up by the Minister of State for Research and Technology has the responsibility to:

- Prepare and formulate policies and programmes for the national development of biotechnology.
- Guide and encourage R & D and the application of biotechnology in Indonesia.
- Enhance the growth of biotechnology networks, both locally and internationally.
- Guide personnel development in biotechnology in Indonesia.
- Guide and encourage the development of bioindustries.

## Biotechnology priorities

In 1988, all the institutions engaged in biotechnology were assessed. These assessments identified the priority areas for biotechnology. These are health, agriculture and industrial processes. This assessment also showed a wide range of states of development, constraints, challenges and opportunities. All the institutes were facing trained personnel shortages and were heavily dependent on public funding. Therefore, activities should be concentrated in prioritized areas and activities with greater chances of success. Based on this, three centres were selected for special attention to enable them to play a major role as growth centres in particular fields:

- The University of Indonesia, Jakarta, for medical biotechnology.
- The CRIFC, Bogor; Ministry of Agriculture; RDCBt, LIPI, for agricultural biotechnology.
- The BPPT, Serpong, Jakarta, for industrial biotechnology.

## Agricultural biotechnology

Indonesia gives high priority to the use of biotechnology for agricultural development. It is expected to play an important role in achieving and maintaining a sustainable agricultural production system. This includes the exploitation and development of biotechnology directed towards increasing productivity, stability, sustainability and equity of agricultural production.

Biotechnology is especially expected to help alleviate production constraints where conventional technology has certain limitations. The intensification of agricultural production in commodities such as rice brings about a build-up of pests and diseases, which constrain production and cause yield instability. Through conventional breeding, breeders face difficulty in obtaining sources of resistance to insect pests, diseases and environmental stresses. It is anticipated that techniques in molecular biology will open up new possibilities for isolating genes that may alleviate these production constraints. In addition, the construction of new genes, such as viral-coat protein genes for protection against plant virus attack, may overcome other production constraints.

In May 1991, CRIFC of AARD organized a workshop in Bogor on 'Potentials of Biotechnology for Enhancing Agricultural Development in Indonesia'. More than 80 people participated from AARD institutes, other government agencies, international aid and agricultural development agencies, universities and private national and international companies, with invited speakers from Indonesia and abroad. The workshop covered a broad range of topics in plant, animal and food production and the use of microbial inoculants. The final sessions involved identification of areas in crop and animal development in Indonesia where biotechnology could play a significant role (Table 4.1). This was recognized as an important initial step in identifying priorities and devising research programmes. These programmes will be initially constrained by shortage of trained staff and equipment. For example, the Biotechnology Laboratory at CRIFC, Bogor, was originally equipped with an emphasis on microbiology and seed technology.

At the workshop it was also decided to establish an Indonesian Society for Agricultural Biotechnology. The Society would organize regular meetings and

**Table 4.1.** Potential impacts of biotechnology on crop and animal production.

| Problem | Biological solution* | Expected output |
|---|---|---|
| **Rice** | | |
| Productivity | Cell and tissue culture, embryo rescue and protoplast fusion | High productivity and quality |
| Insects (stem borer) | Recombinant DNA (use of *Bacillus thuringiensis*) | Resistant variety |
| BLB and blast | Restriction fragment length polymorphism (RFLP) markers | Assistance to breeding programmes |
| Tungro virus | Recombinant DNA (use of viral-coat protein) | Resistant variety |
| Environmental stress | Diagnostic kit | Monoclonal antibodies |
| | Cell and tissue culture | Tolerant variety |
| **Soybean** | | |
| Productivity | *Rhizobium* and mycorrhizae | High productivity |
| Insects (pod borer and sucker) | Recombinant DNA (use of *B. thuringiensis*) | Resistant variety |
| Virus | Resistant variety | Recombinant DNA (use of viral-coat protein) |
| Environmental stress | Diagnostic kit | Monoclonal antibodies |
| | Cell and tissue culture | Tolerant variety |

**Table 4.1.** *Continued*

| Problem | Biological solution* | Expected output |
|---|---|---|
| **Garlic** | | |
| Virus | Clean material | Tissue culture for virus eradication; protoplast fusion; resistant variety |
| Yield improvement | Diagnostic kit | Monoclonal antibodies |
| | Protoplast fusion; superior seeds | |
| **Pepper** | | |
| Virus (CMV) | Control of viral genome by competitive RNA | Vaccine application |
| Lack of resistant traits for CMV | Recombinant DNA | Resistant variety |
| **Potato** | | |
| Planting material | Cell and tissue culture | Rapid multiplication |
| **Citrus** | | |
| Virus (CVPD, tristeza and others) | Shoot-tip grafting | Clean material |
| | Monoclonal antibodies | Diagnostic kit |
| **Banana** | | |
| Stock seed | Cell and tissue culture | Rapid multiplication |
| **Mango** | | |
| Stock seed | Cell and tissue culture | Rapid multiplication |
| **Ornamentals (orchid, chrysanthemum, carnation, Palmae, Anaceae)** | | |
| Seed material | Cell and tissue culture | Rapid multiplication |
| Virus | Monoclonal antibody | Diagnostic kit |
| | Protoplast fusion | Resistant variety |
| Fungi and bacteria | Protoplast fusion | Resistant variety |
| Long-life flower | Protoplast fusion | Superior variety |
| Yield improvement | Protoplast fusion | Superior variety |
| **Oil-palm** | | |
| Planting material not uniform | *In vitro* cloning | Clonal planting material |
| Oil quality low | Protoplast fusion | Better oil quality |
| Parental lines highly heterozygous | Microspore culture | Isogenic lines |
| **Cocoa** | | |
| Planting material not uniform | Cell and tissue culture | Uniform planting material |
| Susceptibility to VSD | *In vitro* selection of resistant cell lines | Clones resistant to VSD |
| Parental lines highly heterozygous | Microspore culture | Isogenic lines |

**Table 4.1.** *Continued*

| Problem | Biological solution* | Expected output |
|---|---|---|
| **Rubber** | | |
| Scion–rootstock incompatibility | Somatic embryogenesis | Complete plant |
| Susceptibility to fungal disease | *In vitro* selection of resistant cell lines | Clone resistant to disease |
| **Coconut** | | |
| Planting material | Cell and tissue culture | Rapid multiplication |
| **Clove** | | |
| Planting material | Cell and tissue culture | Rapid multiplication |
| **Cattle** | | |
| Low population | Embryo micromanipulation (embryo bisection, nuclear cloning) | Increasing population |
| Low protein for consumption | Embryo transfer | Increasing protein production<br>Increasing production offspring |
| Pasciolosis disease | Monoclonal antibodies | Diagnostic kit production<br>Passive immunization |
| Jembrana disease | Monoclonal antibodies | Diagnostic kit production<br>Passive immunization |
| Malignant catarrhal fever disease | Monoclonal antibodies | Diagnostic kit production<br>Passive immunization |
| Low productivity | Recombinant DNA | Increasing milk |
| Low growth rate and late maturity | Recombinant DNA | Speed up growth rate and maturity |
| Low feed conversion | Recombinant DNA of rumen microorganisms | Increasing feed conversion |
| **Poultry** | | |
| New diseases | Monoclonal antibodies | Diagnostic kit production<br>Passive immunization |
| **Fish** | | |
| Low productivity | Recombinant DNA | Increasing productivity<br>Healthy species (low-range health risk) |

* The priorities would be established with the following in mind: (i) the priority must be in line with national agricultural development; and (ii) adoption or development of biotechnological methods to solve the problems would be determined by the availability of existing technology in respect of technical applicability, economic feasibility, social acceptability and biosafety.
CMV, cytomegalovirus.

provide a forum for discussion on research collaboration and promotion of the potential role of biotechnology in different aspects of agricultural production, processing and marketing.

The following priorities for research were tabled:

*Food crops.* It was recommended that research concentrate on rice and soybean, with increases in productivity being achieved through:

- Tissue-culture techniques (somaclonal variation, embryo rescue, protoplast fusion).
- Inoculation of soybean with *Bradyrhizobium* and mycorrhizae.
- Insect resistance through the use of recombinant DNA and plant transformation with the *Bacillus thuringiensis* toxin gene.
- Disease protection through the use of restricted fragment length polymorphism (RFLP) and randomly amplified polymorphic DNA (RAPD) markers in breeding programmes.
- Virus resistance through use of recombinant DNA to transform plants with viral-coat protein gene.
- Diagnosis of virus disease using procedures such as enzyme-linked immunosorbent assay (ELISA) and monoclonal antibodies.

In the late 1990s the priorities changed somewhat, with very little biotechnology research now being done on soybean.

*Horticultural crops.* Garlic, pepper, potato, citrus, banana, mango, ornamentals.

- Tissue culture for virus eradication and rapid multiplication. Protoplast fusion for developing disease-resistant lines.
- Virus protection through use of recombinant DNA viral-coat protein genes.
- Diagnostic kits using antibodies.

*Estate/industrial crops.* Oil-palm, cocoa, rubber, coconut, clove.

- Cloning planting material using tissue culture methods to improve uniformity, rapid multiplication, select disease-resistant lines and improve palm-oil quality.

*Livestock.* Concentrate on cattle, with sheep, buffalo, pigs and poultry being recognized as important.

- Embryo manipulation and artificial insemination in breeding programmes to increase productivity.
- Monoclonal antibodies for disease diagnosis and passive immunization. Vaccine production.
- Rumen microorganism manipulation for increasing feed conversion efficiency.

*Industrial biotechnology*
Emphasis is to be placed on developing bioindustries with a high value-added component, with the research support enabling the industry to become self-sustaining.

- Bioindustries already operating that would benefit from new technology:
  - ethanol fermentation;
  - citric acid fermentation;
  - monosodium glutamate production.
- Bioindustries needed in Indonesia, but no current companies:
  - antibiotic production;
  - steroid production and use of natural products for contraceptives;
  - biopesticides.
- Bioindustries involved in environmental control, particularly control of industrial effluents and waste products.

## Government Activities

The Department of Agriculture through AARD, the Department of Education and Culture through the IUCs, and the Office of the Minister of State for Research and Technology through LIPI and BPPT are committed to supporting agricultural biotechnology R & D.

### Agency for Agricultural Research and Development

AARD is responsible for executing and managing R & D activities within the Ministry of Agriculture. AARD supports the main programme on agricultural biotechnology in Indonesia, comprehensively outlined in AARD (1990).

AARD has developed a reasonably well-equipped Laboratory for Agricultural Biotechnology at CRIFC, now selected as a national centre by the Minister of Research and Technology, which will facilitate and coordinate research activities in biotechnology. The AARD programme involves the CRIFC; the Central Research Institute for Industrial Crops (CRIIC); the Central Research Institute for Horticulture (CRIH); the Biotechnology Research Unit for Estate Crops; the Indonesian Sugar Research Institute; the Central Research Institute for Animal Production; and the Central Research Institute for Fisheries (CRIF). The AARD research facilities, personnel development and research programming were greatly enhanced by a World Bank loan under the National Agricultural Research (NAR) Project (1980–1990) of US$65 million and the Agricultural Research Management Project (ARMP), which followed in 1990 running to 1995, with a World Bank loan of $35.3 million. The government recognizes that biotechnology research is just gaining momentum and needs strengthening. A specific budget allocation has been made for training in biotechnology, research funding, facilities and equipment for the Agricultural Biotechnology Laboratory of CRIFC and the Research Institute for Estate Crops in Bogor and for personnel training.

External funding for AARD having some bearing on biotechnology has also come from the US Agency for International Development (USAID) – Agricultural Research Operational Funding 1986–1990 (US$33 million) and the Applied Agricultural Research Project, Phase II, US$5.99 million (1986–1991), managed by Winrock International. Japanese assistance through JICA has been instrumental in establishing the Agricultural Biotechnology Laboratory at CRIFC, Bogor, and aid from AusAID in building, equipment and training staff for the Research Institute for Animal Production (RIAP), Ciawi (A$33.6 million 1974–1989), and supporting the Research Institute for Animal Diseases (Balitvet/RIAD) in Bogor (A$22.4 million 1980–1990). The UK Department for International Development (DFID) has provided assistance for Balitvet, BORIF and the Sukamandi Research Institute for Food Crops for work on the biological control of pests.

*AARD research programme*

To keep pace with the food needs of an expanding population, to supply raw materials for local industries and to increase the export of agricultural produce, the agricultural biotechnology research programme has the following objectives:

- To assess and apply advanced biotechnologies to accelerate the improvement of food crops, horticultural crops, estate crops, industrial crops, fisheries and livestock production, and to improve processing efficiency and added value of agricultural products through the optimal use of biotechnology.
- To improve dairy cattle production and populations using embryo transfer techniques.
- To improve fish production through genetic improvement and to effectively utilize fish waste for the production of other products, such as proteolytic enzymes.
- To produce diagnostic reagents and vaccines against animal, fish and plant diseases.

*Organization*

The Agricultural Biotechnology Laboratory has been designated by the State Minister for Research and Technology as a National Centre for Agricultural Biotechnology to assess and accelerate the development and application of biotechnology in agriculture, including food and horticultural crops, estate and industrial crops, livestock and fisheries, with close working relations with the IUC at IPB and the RDCBt, LIPI, Cibinong Bogor. It will be the leading laboratory within the institutional network for the application of modern biotechnology and for the diffusion of biotechnological information/skills throughout the members of the network.

This 'pioneer laboratory' will link basic biotechnological skills with the practical areas of AARD's R & D activities. The skills include recombinant DNA technology, enzymology, hybridoma technology and protoplast, cell and tissue-culture technology. The applied areas are animal, fisheries and plant sciences. The skills group will be staffed by junior scientists who are recent postdoctoral fellows and by senior scientists who are knowledgeable in the latest information and have up-to-date technological training. The applied areas will be staffed by experienced senior scientists who have proven records of creativity and productivity, recruited from other AARD institutions.

*Training*

Indonesia has a number of highly qualified scientists working in biochemistry, microbiology, cell biology and genetics. They will form the core of a biotechnology programme. Looking into the future, however, a personnel development programme must be put in place that will ensure both a strong base and continuity.

Scientists will need to be sent abroad for professional meetings, refresher courses, postdoctoral assignments, sabbaticals and similar training to keep them up to date with their peers in the industrial-country laboratories.

Technical assistance is seen as the only way to achieve rapid results. Experts from other countries will be brought to Indonesia to help: (i) train national scientists; (ii) set up laboratories; (iii) determine appropriate research agendas; and (iv) determine the most promising field for investigations. The purpose of this assistance is to make Indonesian scientists self-reliant, to create leadership and to promote interdisciplinary and interlaboratory cooperation. Another important role of the technical team is to design and teach courses in cell biology, microbiology, biochemistry and molecular biology so that the agricultural scientists can learn about the latest developments in modern biological techniques.

*Library and database*

In general, the library facilities in agricultural research institutes and universities in Indonesia are inadequate. AARD established a Central Library for Agricultural Science and Research Communication at Bogor to help overcome this deficiency and to serve all AARD institutes, universities and others. Computer-based, bibliographic searches based on CD-ROM discs have been established for user requests. The Central Library also takes exchange titles and acts as a document centre. Specialized bibliographies are developed by library staff.

## Biotechnology Programmes

Major research activity in different institutions in Indonesia is given in Table 4.2. This section covers current and proposed activities in biotechnology for each of the commodity groupings.

## Food crops biotechnology

Indonesia has made major advances in food crop production, especially for rice and soybean. Continuous and up-to-date research is essential to sustain food production.

High yield always ranks at the top of any farmer's or breeder's list of important traits. Genetically improved crop varieties usually offer the most cost-effective and environmentally friendly means of increasing yields, with or without minimal use of chemical fertilizers and pesticides.

Analysis of factors contributing to such yield gains indicates a key role for disease, pest and stress resistance. Some plant resistances derive from single-gene effects and, if these can be identified and transferred to an advanced cultivar by plant transformation and regeneration, or by crossing – this may involve embryo rescue or the use of RFLPs or RAPDs to track useful genes in a conventional breeding programme – then biotechnology can contribute to the conventional plant breeding approach.

The research strategy acknowledges that there are some shorter-term, 'intermediate technology' approaches, as well as the use of modern biotechnology, which usually requires a long-term programme. Approaches using molecular genetics usually require a team approach. It takes some years to build a successful plant molecular biology team and it will then take further time to start producing useful results. The number of recombinant DNA-transformed plants that are approaching commercial release is small, despite large investments by both private companies and public institutions.

### *Research programmes*

- Genetic improvement of rice and soybean through the use of tissue culture and recombinant DNA genetic engineering.
- Developing effective, nitrogen-fixing inoculants for grain legumes.
- Training scientists and technicians in the use of biotechnology.

The potential use of biotechnology in programmes at the Centre for Agricultural Biotechnology, CRIFC, dealing with rice and soybean production as the priority areas for research, is given in Table 4.3. The research is underpinned by a strong capability in plant tissue culture and microbiology.

### *Outputs*

- Rice and soybean plant regeneration techniques from plant tissue and protoplast cultures.
- Techniques for RFLP mapping, gene isolation, cloning, transformation, embryo rescue, and production of monoclonal antibodies and DNA probes for disease identification.
- Genetically engineered varieties of rice and soybean.

**Table 4.2.** Agricultural biotechnology research activities at Indonesian institutions.

| Institution | Research activities |
|---|---|
| **AARD** | |
| Central Research Institute for Food Crops | Rice: |
| | Somaclonal variation |
| | Anther culture |
| | Protoplast culture |
| | Sweet potato: |
| | Micropropagation |
| | Meristem culture |
| | Rhizobium |
| | Mycorrhiza |
| Central Research Institute for Horticulture | Citrus shoot-tip grafting |
| | Meristem culture of potato and ornamental crops for disease-free planting materials |
| | Garlic: virus eradication |
| Central Research Institute for Industrial Crops | Banana: micropropagation |
| | Clove: micropropagation |
| | Secondary metabolite from *Solanum* sp. |
| Biotechnology Research Unit for Estate Crops | Micropropagation of oil-palm, coffee, tea, rubber and cocoa |
| | Genetic engineering and molecular marker for biotic and abiotic tolerance for major estate crops |
| | Biochemical methods for plant breeding |
| Central Research Institute for Animal Production | Use of microbes to increase straw quality for feed |
| | Embryo transfer |
| | *In vitro* fertilization |
| | Collection of infectious microbes |
| | Vaccine production |
| Central Research Institute for Fisheries | Isolation of proteolytic enzyme from fish waste |
| | Vaccine production |
| **Indonesian Institute of Sciences** | |
| Research and Development Centre for Biotechnology | Embryo transfer |
| | Development of recombinant viral vaccine for Jembrana disease |
| | Genetic characterization of internal thin-tail sheep parasites |
| | Micropropagation of forest tree species and banana |
| | *In vitro* conservation of citrus and yam |
| | Biofertilizer and biopesticides |
| | Isolation of amylase and protease |
| | Bioremediation |
| | Genetic engineering of rice, cassava and forest tree species |
| | Genetic diversity studies of taro (*Colocasia esculenta*) |
| | Isolation of active compound of traditional herbal medicine |

**Table 4.2.** *Continued*

| Institution | Research activities |
|---|---|
| **University** | |
| Inter-University Centre for Biotechnology of Bogor Agricultural University | Use of embryo rescue in soybean-wide hybridization |
| | Somaclonal variation |
| | *In vitro* conservation of ginger, asparagus |
| | Mycorrhiza production |
| | Protease production |
| | Plant-cell and molecular genetics for crop improvement |
| | Analysis of human bacterial pathogens |
| | Genetic mapping of animals |
| | Biofertilizers and biopesticides |
| | Studies on microbial processes and their products |

- Effective and efficient microorganisms for nitrogen fixation and improvement of phosphate absorption.
- Micropropagation techniques for cassava and sweet potato.

*Current status*

The Agricultural Biotechnology Laboratory of CRIFC was inaugurated in 1990. The original emphasis was on microbiology and seed technology and some of the equipment reflects this. Some equipment for molecular genetic work is still needed. The planned development of the Agricultural Biotechnology Laboratory is for a staff of 65 (14 PhD, eight MS degrees), with activity in plant and cell tissue culture, molecular biology, genetics and plant breeding, microbiology, biochemistry, plant protection and plant physiology.

An Australian Centre for International Agricultural Research (ACIAR) project (Improved Diagnosis and Control of Peanut Stripe Virus (PSTV)) started in 1992, linking research in Australia at the University of Queensland, the Queensland Department of Primary Industry, the Commonwealth Scientific and Industrial Research Organization (CSIRO) Division of Biomolecular Engineering, with BORIF and the Central Laboratory for Agricultural Biotechnology. The goal of the project is to protect groundnut from PSTV by transforming groundnut with the coat-protein gene using microprojectiles. A DNA probe for PSTV will be developed for a diagnostic kit for use on field material.

Another ACIAR project linking the New South Wales Department of Agriculture with BORIF and the Central Laboratory for Agricultural Biotechnology investigated the selection of *Rhizobium* inoculant strains for soybean grown after rice and grown in acid soils in newly developed fields in transmigration areas.

Research related to biological control of rice pests is being supported by the European Union (EU) and DFID at BORIF and the Rice Research Institute. This involves a collaborative project with the University of Cardiff and the UK

**Table 4.3.** Potential use of biotechnology in food crop production.

| Problems | Biotechnological solution | Expected output |
|---|---|---|
| **Rice** | | |
| Productivity | Cell and tissue culture, embryo rescue, and protoplast fusion | High productivity and quality |
| Insects (stem borer) | Recombinant DNA (use of *Bacillus thuringiensis* gene) | Resistant variety |
| BLB and Blast | RFLP, RAPD markers | Assistance to breeding programmes |
| Tungro virus | Recombinant DNA (use of viral-coat protein) | Diagnostic kit |
| | Monoclonal antibodies | |
| Environmental stress | Cell and tissue culture | Tolerant variety |
| **Soybean** | | |
| Productivity | *Rhizobium* and mycorrhizal inoculants | High productivity |
| Insects (pod borer and sucker) | Recombinant DNA (use of *B. thuringiensis* gene) | Resistant variety |
| Virus | Recombinant DNA (use of viral-coat protein) | Resistant variety |
| | Monoclonal antibodies | Diagnostic kit |
| Environmental stress | Cell and tissue culture | Tolerant variety |

Natural Resources Institute on brown planthopper parasitoid ecology and acoustic mating signals, pheromones of the white stem borer of rice and soybean pests. The use of the *Metarhyzium* fungus in biological control of the brown and green planthopper of rice is also being investigated. Baculovirus is being used to control the rhinoceros beetle in coconut plantations.

As part of the Rockefeller Rice Biotechnology Network, the Centre for Agricultural Biotechnology (CRIFC) cooperated with the International Rice Research Institute (IRRI) in an attempt to regenerate plants from calluses of javanica rice. This process is an essential prerequisite for any selection based on somaclonal variation, protoplast fusion or plant transformation. Parallel research is studying the natural variation in javanica rice using isozyme analysis.

RDCBt, LIPI, inaugurated in 1986, was designated in 1993 as the second centre of excellence for agricultural biotechnology (Saono, 1995). It has developed transgenic rice with insect resistance traits, with funding from the Indonesian government and the Rockefeller Foundation Rice Biotechnology Network, in collaboration with CRIFC and Dutch counterparts (Leiden and Wageningen Universities). The centre will develop drought- and blast-resistance rice using recombinant DNA and molecular marker technology. An extensive study of genetic diversity of taro (*Collocasia esculenta*) was carried out in collaboration with the Centre de Coopération Internationale en Recherche Agronomique pour le Développement (CIRAD) in France, Wageningen

University, Papua New Guinea, Malaysia and Thailand. A culture collection centre will be developed with initial funding from JICA, Japan.

## Horticultural crops biotechnology

Horticultural research in Indonesia has been reorganized and expanded. The CRIH operates Lembang Research Institute for Horticulture (LEHRI) and Solok Research Institute for Horticulture (SORIF). LEHRI focuses on the production of the highland vegetable crops for Java, including seed production, cultural practices, crop improvement, disease and pest monitoring and control using chemicals and integrated pest management (IPM), meristem culture to free planting material of virus and clonally propagated planting materials. The government of the Netherlands has supported the development of LEHRI's physical facilities and Dutch scientists stationed at LEHRI support work in plant pathology, entomology, virology, pesticide monitoring and integrated plant protection. It is intended to develop an extension programme of IPM to complement the Food and Agriculture Organization (FAO) programme on rice IPM, also supported by the governments of Australia, the Netherlands and the USA.

Biotechnology is being used at LEHRI for the major commercial crops garlic, potato and asparagus. Cabbage and tomato are other important commercial crops being studied. Plant tissue-culture techniques are now being used for production of high-quality (virus-free) planting material, multiplication of superior clones and plant improvement through somaclonal variation. However, the resources, such as trained scientists, budget and research facilities, are limited for this work. Private seed companies are being encouraged to develop collaborative projects with LEHRI, but the commercial seed-producer market is small and a considerable amount of vegetable seed is still imported. A research institute to be devoted to fruit is being developed at Solok, initially focusing on citrus, papaya, mango and banana, the production of disease-free planting stock, varietal improvement, crop production, postharvest activities and marketing.

The research priorities for oil-palm developed at an Agricultural Biotechnology Workshop in Bogor are given in Table 4.4. A possible future research scenario for *Allium* spp. is given in Table 4.5. Depending on the available resources, the research activities will include virus elimination; diagnostic kits for virus identification in plant tissue; use of coat-protein genes to protect against cucumber mosaic virus in pepper; rapid clonal propagation using tissue culture; and development of disease-resistant materials through protoplast fusion. Apart from the virus protection with coat-protein genes, the tissue-culture work would proceed at LEHRI with close interaction on techniques with the Agricultural Biotechnology Laboratory at Bogor. A substantial increase in staffing for biotechnology research is projected for LEHRI. Laboratory buildings and equipment and screenhouse/greenhouse facilities are required before major expansion of activity can occur.

**Table 4.4.** Problems, possible solutions and new technologies for oil-palm.

| Problem | Solution | Technology |
|---|---|---|
| High variability in yield and quality | Clonal palms | Micropropagation |
| High polysaturated fatty acids | Interspecific hybridization | Embryo rescue, protoplast fusion |
| Parental lines highly heterozygous | Development of isogenic lines | Haploid culture |
| Ganoderma trunk rot disease | Hybridization fusion | Cell selection for somaclonal variant, protoplast fusion, gene transfer |

**Table 4.5.** Problems in *Allium* (garlic & shallot) production, possible solutions and the new technologies used.

| Problem | Solution | New technology |
|---|---|---|
| Seed | Micropropagation | Tissue culture |
| Yield increase | Variety improvement | Protoplast fusion, anther culture |
| Diseases | | |
| *Alternaria* | Resistance genes | Gene transfer |
| Viruses | Diagnosis | New diagnostics |
| Environmental stress | | |
| Heat tolerance | Varietal improvement | Tissue culture |
| Soil acidity | | |
| Soil salinity | | |

Research on virus eradication and the clonal propagation of potato is also being conducted at the IUC-IPB.

## Estate crops biotechnology

The Estate Crops Research Institute is funded partly by the government and by state-owned companies and smallholders formed into an Indonesian Association for Research and Development for Estate Crops (Asosiasi Penelitian and Pengembangan Perkebunan (AP3I)). At present there are six autonomous research centres partly funded by AP3I, although they are under AARD: coffee–cocoa, sugar cane, tea, oil-palm, rubber and biotechnology (each has research stations or units). Biotechnology deals with more upstream research, while commodity institutes deal with downstream research. Research institutes for coconut are not under AP3I. Coconut is not regarded as an estate crop and is directly under AARD with other industrial crops. The headquarters institute at Bogor deals with all crops and has a main laboratory, library and management section and a separate field station of 15 ha with a large and well-equipped tissue-culture laboratory. There has been considerable recent support for estate crops research from the government of France, with scientists from CIRAD stationed in Indonesia, with major support for oil-palm biotechnology research at the institute at Marihat. The Deutsche Gesselschaft für Technische Zusammenarbeit (GTZ) has also provided training, equipment and library materials.

Biotechnology research for estate crops started in 1985 in Indonesia with tissue culture and the micropropagation of woody perennials. Oil-palm and coconut are highly heterozygous and seed propagation leads to large plant-to-plant variation. Propagation through seed takes several years and makes for slow progress by conventional breeding. Clonal propagation overcomes this. Practical outcomes of this are most evident for oil-palm, where research on clonal propagation began in 1985 at Marihat. The first clones using somatic embryogenesis from explants produced from young leaf calluses were planted out in 1987 at several locations. Clonally derived plants from superior parent trees are well adapted to the environment, uniform in growth and flower normally, with an increase in female flowers. This leads to increased production of fresh fruit bunches of 16% in the first year and also increased oil extraction compared with plants produced from seedlings. Over 150 different clones have been produced. It takes about 18 months from initial parent tree selection to planting out derived clones in the field. An ambitious programme to produce and market clones commercially was started in the early 1990s. Specialized laboratory facilities have been developed for this, with two researchers and 23 casual workers. Although clonal plants sell at five times the price of seedlings, there is a large market available, with Indonesia catching up with Malaysia in overall annual production of palm oil at 1.3 million t. Problems that may be tackled by biotechnology are given in Table 4.1.

The Biotechnology Research Unit for Estate Crops (BRUEC) laboratory and field station at Bogor will develop the technical knowledge required to support the use of biotechnology in the other research institutes for estate crops, with the overall goal of using tissue-culture techniques to rapidly multiply superior germ-plasm, provide uniform planting material and select genetically superior materials. Five scientists are involved in this work. Biochemical research at BRUEC supports this work in various ways, e.g. developing a rapid method to screen for oil quality in young seedlings before they fruit by correlating leaf-oil components with that in fruits. This involves collaboration with the University of Göttingen (Germany) and CIRAD (France). The programme includes mass propagation of oil-palm, developing molecular markers and recombinant DNA for tolerance to biotic and abiotic stresses, biofertilizers, biopesticides and biofungicides.

Less progress has been made on tissue-culture production of plants for rubber, cocoa, coffee, tea and coconut. For rubber, tissue-culture approaches are being tried to help overcome genetic homogeneity of root stocks. Anther cell culture has resulted in plantlets for two of the 12 commercial clones. For coffee, 150 trees produced by somatic embryogenesis are in the field. At the

Jember Research Institute, a small biotechnology laboratory was started in 1990 to develop micropropagation and tissue culture of coffee. For cocoa, the proembryo stage has been achieved in tissue culture at Bogor. Much of the biotechnology input to cocoa is likely to involve improved methods of fermentation in the postharvest stage. Research on tea has resulted in production of plantlets from leaf callus and through micropropagation of axillary meristems, in collaboration with a Singapore-based commercial biotechnology company.

Coconut is produced mainly by smallholders. Some 3 million ha are under production, with generally low yields. Hybrids have been produced and research to clone these by tissue culture is in progress, but regeneration from tissue culture has proved difficult. There are commercial incentives to develop clones of the Kopyor mutant tree type, which produces a nut with special flesh worth ten to 15 times more than the ordinary nut (see review by Persley, 1992).

### *Sugar cane*

The Indonesian Sugar Research Institute obtains its funding directly from the government and from privately owned sugar-producing companies. Research is undertaken in the following areas:

- Microbial and enzyme biotechnology (starch-hydrolysing enzymes, dextranase, polysaccharide fermentation, amino acid fermentation).
- Biofertilizer production – composting sugar-cane trash; *Rhizobium* inoculants for legume break/green manure crops.
- Environmental biotechnology – control of pollution from sugar-cane factory and ethanol distillery-plant wastes.
- Plant biotechnology – tissue and protoplast cultures, genetic engineering for virus-tolerant and insect-tolerant varieties.

## Industrial crops

The institutes and mandate crops are: the CRIIC (AARD) at Bogor and research institutes at Manado, North Sulawesi, for coconut and other Palmae; research institutes for tobacco and fibre crops at Malang (e.g. cotton, abaca, jute, kapok, ramie, khanaf, tobacco and vegetable oils) and the Research Institute for Spices and Medicinal Crops at Bogor (e.g. pepper, clove, nutmeg, vanilla, ginger, cinnamon, medicinal crops, essential oils, cashews). A large number of crops are involved with varying levels of economic importance and potential. Finding ways to focus biotechnology research with such a large range of species and limited budgets is difficult.

A tissue-culture laboratory was started in 1986 at the Research Institute for Spices and Medicinal Crops at Bogor. The main emphasis has been on regeneration of plants from callus, micropropagation to produce clonal material, with some research on secondary metabolites in tissue culture and germ-plasm preservation of rare medicinal plants by tissue culture. Tissue culture of seven medicinal-crop species, six essential-oil species, two spices, two fibre crops and five other industrial crops has been attempted. Research on mass propagation of banana is being undertaken, sponsored by a private company. The large levels of polyphenols occurring in tissue cultures of some crops has made plantlet generation difficult. Collaboration with an established tissue-culture laboratory in an industrial country would be beneficial. Some major problems in production of these crops may be amenable to biotechnology approaches. For example, the annual death of millions of clove plants is thought to be due to a xylem-like bacterium (XLB). Tissue-culture somaclonal variation may help develop resistant lines. Genetic variation in clove, nutmeg and coconut may be used to advantage by cloning superior material.

## Animal production

Livestock and poultry production has steadily increased over the last 10 years but has not kept pace with demand. Animal nutrition is a key factor in increasing productivity per animal. Private companies have developed concentrate feeds for intensive production systems for poultry and pigs. Large-volume feed stocks, such as cassava chips, which could be readily produced in Indonesia, are underutilized. Demand for milk products has increased, prompting an interest in animal breeding, embryo manipulation and the utilization of exotic genes as a potential answer to increased efficiency of production. However, the greatest challenge, although perhaps not the most exciting approach, is to reliably produce amounts of feed with sufficient protein. The utilization of forage from trees such as *Gliricidia sepium* and *Calliandra callothyrsus* used in agroforestry production systems is still under study. There is considerable promise in this work, particularly in the effect of polyphenols on protein sparing in the rumen and the effect of treatments after cutting and feed-mixture composition on digestibility and efficiency of utilization of protein.

Biotechnology can play a role in these issues, e.g. by fermentation of by-products to produce feeds and feed supplements, study of feed utilization, disease diagnosis through use of antibodies in ELISA tests and DNA probes, disease control through vaccination and manipulation of breeding cycles and embryos.

The Central Research Institute for Animal Sciences (CRIAS) of AARD coordinates RIAP at Ciawi and RIAD at Bogor. Both have several field stations. The Directorate General of Livestock and the Department of Cooperatives manage animal distribution and marketing schemes. The Village Cooperative Units (KUD) have played an important part in the increase in dairy production. Their experience in adaptation of exotic dairy animals in the tropics and a link between research scientists and village production systems would provide valuable insights into the real problems and issues in animal production.

Research is also conducted by the Central Research Institute for Biotechnology of LIPI, and the IUC and Faculty of Veterinary Medicine at Bogor Agricultural University.

RIAP and RIAD have benefited from a large amount of support from the Australian government in building and equipping the institutes, training scientists and establishing research programmes. At RIAD a 10-year programme supported staff development and equipment. Both institutes had Australian consultants over a 10-year period. A part of the research involved biotechnology. For RIAP, the research programmes centre around: animal feeds, developing ration mixtures and forage additives; pasture and tree-forage production and testing their feed qualities *in vitro* and *in vivo*; fermentation of cassava and by-products, such as rice straw and molasses, to produce feed supplements; monitoring fertility, manipulation of ovulation and artificial insemination to increase reproduction efficiency; embryo-transfer techniques; and selection for twinning in sheep.

Development of vaccines and antibodies for disease-diagnostic purposes is an important research activity for RIAD. Current research, supported by an ACIAR collaborative project, involves use of ELISA serological techniques (both polyclonal and monoclonal antibodies) to diagnose disease, e.g. haemorrhagic septicaemia in cattle; anthrax in cattle, sheep and goats; brucellosis in cattle; heptospirosis in dairy cattle and pigs; Newcastle disease in poultry; neonatal scours in pigs. Monoclonal antibody production is in progress as part of this project. DFID is supporting a project to identify trypanosomosis based on RFLPs, to develop a DNA probe and amplification of the reaction with the polymerase chain reaction (PCR) technique. Other research includes developing and testing vaccines for Newcastle disease (with ACIAR collaboration), vaccines for other viral diseases and neonatal scours in pigs caused by *Escherichia coli*.

ACIAR has also provided funding for RDCBt, LIPI, to carry out genetic and immunological characterization of high resistance to internal parasites in Indonesian thin-tail sheep and production of recombinant viral protein vaccine for the control of Jembrana disease in Indonesia.

## Fisheries

CRIF (AARD) administers research institutes for marine fisheries, freshwater fisheries and coastal aquaculture. In the Research Institute for Coastal Aquaculture, research covers: development of feeding materials for shrimp and fish; improved genetic stock; disease diagnosis; production of enzymes, such as protease, from fish waste; use of proteases to descale fish; and fish preservation through drying and salting. Production of proteases by *Bacillus stearothermophilus* is being investigated at RIAP. Indonesia has great potential to expand fresh and coastal fisheries and shrimp production.

## Food

Biotechnology has been used in food production in Indonesia for centuries as solid or liquid fermentations to produce *tempe*, soy sauce and *tauco* (soybean), *oncom* (pressed groundnut cake) and *tape* (fermented cassava or glutinous rice), mostly at the household or small-scale manufacturer level (there are estimated to be more than 40,000 *tempe* manufacturers). It is difficult to scale up these industries and, although institutes such as the Central Research Institute for Applied Chemistry, LIPI, at Bandung have developed improved inoculants for *tempe* production, the market is small (currently about 300 kg $year^{-1}$) and likely to remain so, simply because the cost of distributing inoculants to individual households would be out of all proportion to the product value. The production of these items is marked by low inputs, small investments, low added value and an unskilled labour base, with larger factories producing up to 1.5 t *tempe* $day^{-1}$.

Mushroom production is a feasible activity with considerable potential for improvement through biotechnology, but there is so far little research on production.

Medium-scale industries applying biotechnology are distilleries manufacturing ethanol from molasses, beer breweries, citric acid production from solid cassava waste and monosodium glutamate production using molasses as substrate. These industries are characterized by higher investment, higher levels of skills and higher added value.

Past R & D in the country has been dominated by efforts to improve the traditional fermentation processes through standardization of raw materials; isolation, identification and characterization of traditional inocula; and studies of factors influencing fermentation. The biotechnology activity at present is mainly research. There are some pilot plant development facilities available at the Food Technology Development Centre of IPB at Bogor and at ITB in Bandung. Contract work has mainly been in the field of food processing and improvement of local village-scale fermented food products, such as *tempe* and soy sauce. Use of new substrates, such as cassava residues, for ethanol production and development of food colouring agents from fungi are under study in the Laboratory for Microbiology and Fermentation Technology, ITB.

## Forestry

There is little biotechnology research within the Ministry of Forestry. Research on tissue culture and micropropagation of forest tree species is taking place at the RUD Centre for Biotechnology, LIPI and at BIOTROP, Bogor and on *Rhizobium* and mycorrhizal inoculants at the IUC-IPB, IUC-UGM and BIOTROP. Commercial companies developing new plantations are also conducting their own R & D in these areas.

### Germ-plasm conservation

To support the biotechnology work at CRIFC, CRIH, CRIIC and other AARD research institutes, LIPI and the IUCs, LIPI has developed a germ-plasm collection for conservation of the genetic resources of Indonesian plants. This involves germ-plasm collection, description, storage, documentation and distribution and is a vital component of the national programme on agricultural biotechnology. The Minister of Agriculture recently set up the National Commission for the Preservation of Plant and Animal Genetic Resources.

## Private-sector Activities

In the Indonesian private sector there is interest in biotechnology, but little current R & D activity. There is a long tradition of fermented foods in the country, but these are mostly produced by small manufacturers without the capital resources to move into medium- or large-scale processes. There are some medium-scale industries involved in alcohol production from molasses, beer brewing, citric acid production from cassava waste, soy-sauce production and monosodium glutamate and single-cell protein production.

PT Indah Kiat (Riau, Pakabaru) are using biotechnology (tissue culture, recombinant DNA and molecular marker) for tree improvement in collaboration with the University of Peking. PT Monfori Nusantara (Parung, Bogor) are obtaining high-value species of teak (*Tectona grandis*) and *meranti* (*Shorea* species) using DNA marker technology.

Animal vaccines are produced commercially by PUSVETMA, a government-owned private company that undertakes some research, as does Perum BioFarma (Pasteur Institute, Bandung), the parallel company producing a wide range of vaccines, antisera and diagnostics for human public health services.

PT Kalbe Farma, a large, semi-government-owned pharmaceutical company, has developed a division of plant biotechnology looking at tissue culture of horticultural crops, potato and strawberry.

## University Activities

### Inter-University Centres

The establishment by the Ministry of Education in 1985 of three IUCs for Biotechnology was financed with World Bank loans of US$23 million. Their role is to train faculty members from other universities, operate a postgraduate degree programme, conduct focused research and seek private-industry links in up-to-date research and technology in the field of biotechnology.

### *IUC for agricultural biotechnology, IPB, Bogor, Bandung (now Research Centre for Biotechnology)*

The Centre moved into a new building on the Dermaga campus in 1991. It has laboratories equipped for tissue culture, microbiology, molecular genetics, fermentation and other aspects of biotechnology. Work is being done on *Rhizobium* and mycorrhiza inoculants, tissue culture of potato and other crops, cellulase production by *Trichoderma*, fermentation of agricultural by-products and waste-water treatment.

IUC-IPB has the responsibility for postgraduate training in agricultural biotechnology in Indonesia. It has instigated a 'sandwich' course, whereby Indonesian students are sent overseas for part of their postgraduate studies, but work on a research topic of relevance to Indonesia. Their overseas adviser is also invited to be a visiting scholar at IUC-IPB so that the adviser is able to cosupervise the Indonesian part of the thesis and to lecture to a broader range of students at the IPB campus.

The present programme includes plant molecular genetics for crop improvement, analysis of human bacterial pathogens, genetic mapping of animals, biofertilizers and biopesticides, and studies on microbial processes and their products.

### *IUC for industrial biotechnology, ITB (now Inter-University Research Centre)*

This IUC moved to spacious new premises in 1990 and US$4 million worth of equipment is now in place, ranging from pilot-plant fermentation, waste-water treatment and enzyme preparation facilities to laboratories for microbiology, enzymology, molecular genetics, biochemistry and analytical chemistry. Most of the IUC scientific staff have received or are receiving overseas training leading to higher degrees. Members of the IUC staff hold other permanent faculty appointments. There are cooperative research projects with Kyoto University, Japan; Berlin University, through GTZ; the universities of Strathclyde (on industrial waste-water treatment) and Kent (International Institute for Biotechnology) in the UK, supported by DFID.

The IUC has three major areas of activity: microbiology and fermentation; enzyme technology; and waste-water treatment.

The microbiology area covers penicillin, high-fructose syrup and methanol production. A molecular genetic laboratory for DNA sequencing and transformation is being established with assistance from Kroningen State University Biology Centre, Haaren, the Netherlands. DNA transformation systems for *Streptomyces* and *Bacillus megatherium* are being developed. The enzyme technology group is examining high-fructose syrup and penicillin production. Treatment of waste-water from palm-oil production is at the pilot-plant testing phase. Textile production waste-water is being treated by a contact reactor and adsorption on activated carbon. A large-scale, river-flow simulation apparatus is being used to examine the kinetics of microbiological, microfaunal and chemical changes when various pollutants, such as phenol and textile dyes, are added. Also under way are a vaccine for screw-worm fly, biodegradation of liquid effluents and the molecular pathogenesis of *Salmonella typhi*.

*IUC at Gadjah Mada University, Yogyakarta*

Although this was established as a Centre for Medical Biotechnology (e.g. vaccine production for tropical diseases) there is also ongoing research related to agriculture – with selection of bioinsecticides (*B. thuringiensis* strains for grub control in forestry) in collaboration with ITB, where the *B. thuringiensis* is multiplied in fermenters, research on mycorrhiza for agroforestry and forest tree species and *Rhizobium* strains for grain legumes.

## Government, University and Industry Cooperation

The government is promoting links between industry – government and/or private companies – and the government research institutions. This is happening in various ways:

- Establishment of research programmes and research networks for biotechnology, which are undertaking projects whose outcomes will be methodology, materials and/or finished products that could be directly useful to industry (e.g. penicillin production, waste-water treatment, genetically improved seed, or planting material (oil-palms), *Rhizobium* and mycorrhizal inoculant strains for crops in acid soils, superior *B. thuringiensis* bioinsecticide strains and *B. thuringiensis* toxin genes).
- Encouraging industry sponsorship of research, either in collaboration with the research institute or on a fee-for-service, turnkey basis. The incentive for this is the ability of the institute to retain any profits from royalties and the fees paid for the research.
- Supporting the commercial production of large amounts of oil-palm planting material regenerated through tissue culture and micropropagation at the Central Research Institute for Estate Crops at Marihat. When industry sees genuine production gains from superior germ-plasm, it will be prepared to pay the full cost for the materials.
- Two projects funded by USAID and World Bank loans link Indonesian universities with AARD research programmes. This is done through a grants scheme that enables researchers from both systems to collaborate in the commodity crop research programmes on rice, maize and soybean, including biotechnology. This is administered by CRIFC, but with a joint AARD/university selection and review committee.
- In 1994 an initiative to have a network between universities and research institutes resulted in the establishment of the Indonesian Biotechnology Consortium (IBC). It started with only seven IUC centres but now has 27 members, including universities, government institutions and semi-private and private companies.

## Training

An ambitious programme of training scientists in biotechnology-related disciplines, both abroad and in Indonesia, is under way. Every biotechnology institute has a large number of its staff currently out for training or retraining, many undertaking higher degrees. This is scheduled to continue for some years. The IUCs have a specific mandate to train staff and students from other Indonesian universities; also the IUC at IPB is developing a large postgraduate programme involving course work and research specifically orientated to agricultural biotechnology.

The Ministry of Research and Technology, with a loan from the World Bank, now has a major programme to train scientists in different fields, including biotechnology.

Secondment of staff from industrial-country laboratories to Indonesian universities and research institutes for long periods to work with Indonesian scientists is recognized as important (AARD, 1990).

From 1989 to 1997, the number of PhDs in biotechnology increased from 50 to 102 and MS and BSc degree holders from 75 to 247 (Falconi, 1999).

## Regulatory Procedures

Guidelines have been developed for genetically engineered organisms and their products, and the issue is under discussion by the National Committee on Biotechnology. Biological control agents, vaccines and microbial inoculants may be covered adequately by existing legislation and testing procedures.

Indonesian biosafety regulations for release of genetically modified organisms (GMOs) were put in place in 1997. The Minister of Agriculture released a Ministerial Decree for Genetically Engineered Agricultural Biotechnology Products in 1997. To implement the decree, a committee for biosafety was formed in 1997, which is supported by a technical team consisting of experts in plant biotechnology from national institutes and universities. The technical team formulated a series of guidelines for release of genetically engineered organisms. They include the general guidelines for plant, cattle, fish and microbes and the specific guidelines for each item. The 1997 decree did not cover food safety. To fulfil this need, another decree was released in 1999 as a collective decree for four ministries (Ministry of Agriculture, Ministry of Estate Crops and Forestry, Ministry of Food and Ministry of Health) for Biosafety and Food Safety of Genetically Engineered Agricultural Biotechnology Products. The number of committee members and technical team members was expanded to represent different parties. The guidelines on food safety of GMO products have been drafted (Slamet-Loedin *et al.*, 2000).

Indonesia enacted a patent law in 1989, which came into force in 1991. In that patent law, in respect of biotechnological invention, no patent could be granted for any process for production of food, drinks for human and animal consumption or for newly introduced plant varieties, animals or their products. This patent law was revised in 1997 (UURI no. 13) in accordance with World Trade Organization (WTO) regulations and Trade-Related Aspects of Intellectual Property Rights (TRIPs) and those articles were deleted, while patent protection was extended from 14 to 20 years (Saono, 1995). The Plant Variety Protection Act was accepted and became law in 2000. Protection can be granted to a known plant variety with additional gene/genes introduced through genetic engineering.

Current customs and import regulations can adversely affect the import of perishable consumables, such as radioactive chemicals, enzymes and specialized chemicals needed for molecular biology work. These can deteriorate rapidly in storage in customs before release (sometimes this can take up to 3 months).

Regulations regarding the export and use of Indonesian germ-plasm outside Indonesia are currently under consideration. Ways to obtain some form of return for its use abroad are being sought. The 'Farmers' rights' approach is one possibility.

## International Cooperation

Advances in biotechnology are taking place at such a rate internationally that close liaison among scientists working at the leading edges of biotechnology is vitally important. Funds for travel to international congresses and symposia are of special importance for scientists, as is access to international research data banks. Being able to access current literature is a high priority and sufficient funds should be assured to key AARD libraries so that they can maintain subscriptions to important periodicals, or at least maintain a complete run of journals at the Central Library for Agricultural Science, Bogor, or obtain copies of reprints from abroad.

The ACIAR model of collaboration between laboratories in Australia and Indonesia on projects of mutual benefit to both countries has been highly productive in both scientific achievements and technology transfer. In this model scientists from both countries may spend time working in each other's laboratory and there are several visits from the Australian scientist(s) to Indonesia each year. Operating, travel and some capital equipment funds are supplied for the project.

While developing biotechnology skills, scientists must keep in mind that biotechnology is not an independent discipline, but a set of new techniques and tools for use by biologists in their R & D efforts.

Indonesia, as an active member of ASEAN, has participated in COST meetings concerning cooperation on biotechnological development (since 1982). It is also an active member of ASEAN–Australia projects on biotechnology.

BIOTROP is an initiative of the South-East Asian Ministers of Education Organization (SEAMEO). Its function is to: undertake critical, problem-solving research in natural resources conservation; train regional scientists in tropical biology; disseminate information on BIOTROP's research findings; and foster international and national cooperation and information exchange. BIOTROP has two programmes that involve biotechnology – tropical forest biology and agricultural pest biology – with a small IPM component.

## Conclusions

Indonesia has placed a high priority on the development of biotechnology. Seven major centres for research in biotechnology have been developed since 1985 and another is partly completed. Also, biotechnology is being utilized in a major way at another seven institutes under AARD management, at BIOTROP and by private companies. The policy framework for biotechnology in Indonesia has been carefully developed for the year 2000 and beyond. A major training programme within Indonesia and abroad was initiated and is still in progress, with some institutes having as much as 50% of their staff currently involved.

Research activities are at different stages of development for various aspects of biotechnology. A start has been made in all the major areas of interest. As in the rest of the world, plant and microbial biotechnology is more strongly developed than animal biotechnology. Plant tissue-culture and micropropagation techniques are well established at several laboratories and large-scale commercial production of oil-palm planting material was developed at the Research Institute for Oil Palm at Marihat. Similar developments for some other plantation crops are likely within a few years. The use of tissue culture to eradicate virus is also likely to support commercial production of planting material of potato and several horticultural crops in the near future, and selected strains of *Rhizobium* and mycorrhiza are available for commercial inoculant production. Techniques in the field of molecular biology, including molecular genetics and the use of recombinant DNA or genetic engineering, are being established in Indonesia but it will be several years before they make a significant impact on research outcomes relevant to increasing agricultural production. A Workshop on Agricultural Biotechnology at Bogor started the process of defining production problems where biotechnology could make a significant impact (Tables 4.2–4.5). The workshop also emphasized the importance of coordinating research to reduce duplication and regular meetings for exchange of experiences.

In animal biotechnology much emphasis has been placed on manipulating reproduction, through control of ovulation, artificial insemination and embryo transfer in cattle and sheep. There is a long history of research on animal disease diagnosis and control through vaccination in Indonesia. The use of modern biotechnology in these endeavours has just begun, with the use of poly- and monoclonal antibodies in ELISA diagnostic kits and the development of DNA probes for trypanosomes that cause disease in cattle. Inadequate nutrition, however, is the major constraint to most animal production and, although some research is being undertaken in this area at RIAP, its importance warrants a much greater emphasis.

There are several constraints on the development of biotechnology. Physical facilities are generally very good and major items of equipment are available. There is a need, however, for a modest capital equipment budget to fill in the gaps, usually for small, low-cost but nevertheless vital pieces of equipment. These items are usually identified when scientists experienced in the field actually start the laboratory work programme.

Continuity of operational funding is a problem. Long-term research such as biotechnology needs a commitment over several years. Molecular biology involves expensive consumables. Library funds have been severely constrained. Much more attention needs to be given to selecting and maintaining a core set of relevant journals and books at least at one centre in Indonesia, such as the Central Agricultural Library at Bogor, with concerted efforts made to supply articles to scientists in the research institutes. There needs to be a recognition that collection and maintenance of germ-plasm resources is a key resource for biotechnology. It is urgent for national collections for conservation, evaluation, utilization and documentation of collected germ-plasm to be further improved in the near future, before the material is lost through land development and changes in agricultural practice.

The 'intermediate' biotechnologies have an important role to play and need much more emphasis in national planning. Such intermediate biotechnologies are crop improvement; IPM; biological control of weeds, fungi and insect and nematode pests of plants; inoculants beneficial in plant nutrition; utilization of rice straw.

The basic biological science disciplines underpin the use of biotechnology (e.g. plant pathology, biochemistry, physiology, entomology, genetics and breeding, microbiology) and attention should be given to maintaining focused research in these areas by appropriate project development, evaluation and funding.

There is considerable potential gain for Indonesia from further development of agricultural biotechnology. The returns on investment will be substantial in research to provide planting materials for agricultural, plantation, industrial and horticultural crops that are uniform, high-yielding, disease-free, pest- and stress-tolerant, and improved nutrition and germ lines for animals and aquaculture. However, such research needs to be carefully focused on getting the end-product to the market-place. This will involve careful planning and testing of scale-up procedures and market assessment. Intermediate biotechnologies for IPM and biological control of plant pests are likely to have a high return on investment, and Indonesia is one of the leading countries in the world in adoption of IPM. The benefits from this approach need to be carefully documented to assess the role it should play in national priorities.

The development of disease-diagnostic kits, particularly for animals and agricultural crops (e.g. tungro virus detection in rice), has considerable potential benefit. Use of RFLP and RAPD methods in plant breeding should also enable a much more focused programme of selection for disease resistance involving quantitative traits.

For agricultural crops, biotechnology should also play an important role in breeding for disease resistance through embryo rescue and protoplast fusion. A longer-term prospect utilizing recombinant DNA technology is the development of resistance to virus (coat-protein protection) and insects through transforming the *B. thuringiensis* genes into plants, if the level of expression of the genes can be increased sufficiently to give enough insect death in the field.

The better management of fisheries resources through biotechnology, e.g. better genetic base, improved nutrition and disease diagnosis and control, also has considerable potential.

For bioindustries, the development of improved fermentation and formulation techniques for a range of products and enzyme products to produce high-value materials, such as high-fructose syrup, from low-value materials, such as starches, also has potential for creating new commercial products. In the related medical field, production of antibiotics also appears to be a promising commercial prospect.

Although the level of national funding is increasing, we should recognize the important role outside funding agencies have played and will continue to play in the development of biotechnology in Indonesia. The provision of capital equipment, including buildings, has in large part been supported in this way. Support to ensure adequate operational costs may also be necessary and, for bilateral agencies, would best be achieved by support for particular projects. It must be realized that it will be a lengthy period (5–10 years minimum) before many of the biotechnology projects produce products and this needs to be acknowledged by the respective agencies in developing appropriate project support.

The most valuable help bilateral agencies can offer initially is support for scientists from industrial countries to collaborate closely with Indonesian scientists, which is best done on an institution-to-institution basis for particular projects. Initially this may mean supporting industrial-country scientists working in Indonesian laboratories for periods of at least 2 years. The institute-twinning approach enables continuity of collective memory if a particular scientist can spend only a limited time in Indonesia.

## References

AARD (1990) *Agricultural Biotechnology Research Programme in Indonesia.* Agency for Agricultural Research and Development, Jakarta, 76 pp.

Collective Decree of Ministry of Agriculture, Ministry of Estate Crops and Forestry, Ministry of Health, Ministry of Food No. 998.1/Kpts/OT.210/9/99; 790.a/Kpts-IX/1999; 1145A/MENKES/SKB/IX/1999; 015A/Nmeneg PHOR/09/1999.

Falconi, C.A. (1999) *Agricultural Biotechnology Research Capacity in Four Developing Countries.* ISNAR Briefing Paper 42, The Hague, The Netherlands.

Manwan, I., Moeljopawiro, S. and Mariska, I. (1990) Present status of agricultural biotechnology application in Indonesia. Paper presented to a Workshop on Recent Advances in Biotechnology, Jakarta, November 1990.

Moeljopawiro, S. (2000) Indonesia. In: Tzotzos, G.T. and Skryabin, K.G. (eds) *Biotechnology in the Developing World and Countries in Economic Transition.* CAB International, Wallingford, UK, pp. 88–91.

Persley, G.J. (1992) *Replanting the Tree of Life: Towards an International Agenda for Coconut Palm Research.* CAB International, Wallingford, UK, 145 pp.

Research Highlights and Research Reviews (1991) *The Strengthening of Pioneering Research for Palawija Crops Production Project.* ATA-378 Project, Bogor Research Institute for Food Crops and Japan International Cooperation Agency (JICA).

Saono, S. (1995) Biotechnology in the member states of ASEAN. *Biotechnology* 12, 400–417. VCH.

Slamet-Loedin, I.H., Toruan-Mathius, N., Sukara, E. and Larossi, A.T. (2000) Status of R & D on transgenic plants in Indonesia. Paper presented in ASEAN–China Workshop on Transgenic Plants, Beijing, China.

# Pakistan

5

Yusuf Zafar

| | |
|---|---|
| Area ($km^2$) | 803,940 |
| Cropland | 27% |
| Irrigated cropland | 171 $km^2$ |
| Permanent pasture | 6% |
| Population | 141.5m |
| Population per $km^2$ | 56 |
| Ann. pop. growth rate (2000 est.) | 2.17% |
| Life expectancy (men) | 60.2 yrs |
| (women) | 61.9 yrs |
| Adult literacy | 37.8% |
| GDP (1999 est.) | US$282 bn |
| GDP per head | US$2000 |
| Growth in real GDP (1999 est.) | 3.1% |
| Inflation (1999) | 6% |
| Agriculture as % of GDP | 25% |
| Agriculture and food processing as % of manufacturing sector | 41.0% |
| Value of agricultural exports (1998) | US$660m |
| Agricultural products as % of total exports | 33% |

Major export commodities: cotton, rice, other agricultural products

Major subsistence commodities: wheat, rice, sugar cane, fruits, vegetables

## Summary

The potential of biotechnology in Pakistan was formally recognized in 1981, when a course on recombinant DNA technology was organized by the Nuclear Institute for Agriculture and Biology (NIAB), Faisalabad. NIAB is one of three agricultural centres of the Pakistan Atomic Energy Commission. The training workshop asked the government of Pakistan to develop a national centre of biotechnology and genetic engineering. The Ministry of Education later approved the creation of a Centre of Excellence in Molecular Biology (CEMB), to be built on the campus of Punjab University.

The National Institute for Biotechnology and Genetic Engineering (NIBGE) was approved in 1986. The aims of the institute are to develop, adopt and apply innovative modern research in agriculture, industry, health and the environment.

Agriculture is the backbone of Pakistan's economy, contributing 25% to gross domestic product (GDP). About 68% of the people depend on agriculture for their livelihood, engaging about 50% of the country's workforce.

## Introduction

Agricultural biotechnology commenced in Pakistan with a training course in recombinant DNA in 1981 (Table 5.1). Traditional biotechnology (tissue culture and biofertilizer) has been carried out at many of the research centres for some time (Khan, 1997; Khan and Afzal, 1997), although modern biotechnological research is restricted to only two of the centres: the Centre of Excellence in Molecular Biology and the National Institute for Biotechnology and Genetic Engineering (NIBGE). These élite centres, however, have some major weaknesses. Although the centres do have the scientific capabilities to use the most modern biotechnological tools, the research effort is still comparatively small. The centres still lack a sufficient number of researchers and adequate financial resources to mount large-scale research efforts, though this has been achieved for potato.

Recent developments in plant biotechnology have greatly increased the possibility of crop improvement, allowing the manipulation of genetic material with greater accuracy in a much shorter time frame than is possible using conventional breeding methods.

Research in Pakistan agricultural biotechnology is divided into two broad categories:

- Traditional biotechnology
  - Fermentation – biopesticides, biofertilizers.
  - Tissue culture – mass production of disease-free plants.
- Modern biotechnology
  - Molecular breeding – DNA fingerprinting.
  - Genetic engineering – development of crops with novel traits.

**Table 5.1.** Development of biotechnology in Pakistan.

| | | | |
|---|---|---|---|
| Recombinant DNA Technology Course | 1981 | USNS/PAEC/GOP | NIAB, Faisalabad |
| UNIDO Review Mission for ICGEB | 1982–1984 | UNIDO/GOP | |
| Establishment of Centre of Excellence on Molecular Biology (CEMB) | 1984 | Ministry of Education (MOE) | Punjab University of Lahore |
| S & T Fellowship Scheme | 1985–1995 | MOST | 1000 PhDs (Biotechnology 3%) |
| National Institute for Biotechnology and Genetic Engineering (NIBGE) | 1986 approved | GOP | Faisalabad |
| Centre of Chemistry and Biotechnology, Agricultural Biotechnology | 1995 | UGC | University of Agriculture, Faisalabad |
| Institute of Biochemistry and Biotechnology | 1995 | UGC | Punjab University, Lahore |
| MPhil Biotechnology | 1995 | QAU/NIBGE | NIBGE, Faisalabad |
| MSc Biotechnology | 1996 | UGC | Karachi University, Karachi |

GOP, Government of Pakistan; NIAB, Nuclear Institute for Agriculture and Biology; UNIDO, UN Industrial Development Organization.

## Traditional Agricultural Biotechnology

### Biopesticides

The increasing use of chemical pesticides has become a growing concern in recent years, with indications that their use will increase and intensify in the short term. It is therefore urgent that we develop and promote the use of alternative methods of crop protection. It is particularly important that efforts be made to substitute chemical pesticides with biopesticides, which are environmentally friendly.

The two most important advantages of biopesticides are: (i) they are target-specific and do not harm beneficial organisms; and (ii) they do not leave harmful residues. The following are some of the important biopesticides:

- Trichogramma (egg parasitoid) to control lepidopteran pests, such as sugar cane internode borer.
- Fungi (*Trichoderma* and *Gliocladium*) to control root rot and wilt disease in pulses.
- Baculoviruses.

  Nuclear polyhedrosis virus (NPV) of *Heliothis armigera* for cotton, oilseeds, pulses, vegetables and millets.

**Table 5.2.** Biopesticides programmes in Pakistan.

| Centre | Mechanism | Crops | Status |
|---|---|---|---|
| AEARC | Control of pests by parasitoids/predators | Sugar cane/cotton | Small commercial venture |
| IIBC, Rawalpindi | Control of pests by parasitoids/predators | Sugar cane/cotton | Small commercial venture |
| ITAR, Karachi (PARC) | Neem-based formulation | Cotton | STADEC/PCSIR Nimolene |
| CAMB, Lahore | Novel Bt biopesticide | Cotton/rice | Lab scale |
| CAMB, Lahore | Fungi-based pesticide | Cotton | Lab scale |

Bt, *Bacillus thuringiensis.*

NPV of tobacco caterpillar (*Spodoptera litura*) for tobacco and cotton. Granulosis virus (GV) for sugar-cane internode borer.

- *Bacillus thuringiensis.*
- Neem.

The programmes at research centres in Pakistan are summarized in Table 5.2.

## Biofertilizers

The potential of certain microorganisms to improve the availability of nutrients to crop plants has long been known. In view of the rise in the cost of chemical fertilizers and their adverse effects on the environment, these microorganisms (collectively called biofertilizers) have become increasingly important. They are considered to be particularly important in tropical countries, where soils are often deficient in organic matter and essential plant nutrients, due to high temperatures and intense microbial activity.

Most biofertilizers fix atmospheric nitrogen to ammonia by a complex metabolic process. There are two types of biofertilizer: symbiotic and free-living. The former, which require symbiotic association with plants, are represented by *Rhizobium*. The latter, which can fix nitrogen independently, include *Azotobacter, Azospirillium*, blue-green algae (BGA) and *Azolla.*

## Tissue culture

Modest tissue culture facilities were developed as early as 1968 in the Botany Department of Peshwar University. There is enormous scope for mass propagation of disease-free plants in several important vegetatively grown crops (sugar cane, potato, banana, date-palm). None of these have acquired a commercial level of production except potato. Potato production with 5% annual growth is a real success story, mainly due to the extensive involvement of the Pak-Swiss Potato Project (Table 5.3). This venture has now been taken up by several small-scale private firms and non-governmental organizations (NGOs).

**Table 5.3.** Agricultural biotechnology research in Pakistan.

| Centre funds | Area | Researchers/ PhD | Facilities | Commercial product | Research |
|---|---|---|---|---|---|
| AARI, Faisalabad | Tissue culture of sugar cane, potato, wheat | 12 | Poor | Virus-free potato tubers | Punjab government/ Pak-Swiss Project |
| ARI, Tandojam | Tissue culture of banana | 5 | Poor | Nil | Sindh government ADB/ARP-II |
| ARI, Tarnab | Tissue culture of fruit crops | 3/1 | Poor | Nil | NWFP government |
| ARI, Sariab, Quetta | Tissue culture of dry fruits, trees | 3 | Poor | Nil | FAO/IDB/KRF |

ADB, Asian Development Bank; FAO, Food and Agriculture Organization.

# Modern Agricultural Biotechnology

## Molecular breeding

DNA fingerprinting is a powerful technique that is now widely used in conventional breeding programmes. This technology is extremely useful in the following areas:

- To evaluate the genetic diversity of a specific crop.
- To determine the relationships among species.
- To link some molecular markers with a specific trait to speed up the breeding process.
- To isolate genes of interest by making dense genome maps.

In Pakistan, the Cotton Genome Project was initiated in 1996. It is the only project of its kind in the country.

## Genetic engineering for crop improvement

Genetic engineering of plants began in the biotechnology group of NIBGE in 1986 (Table 5.4). No transgenic plant has yet been released commercially in the country, either developed through local efforts or imported from industrial countries.

Use of biotechnology in crop improvement, therefore, is comparatively new in Pakistan. Most of the work is concentrated on chickpea, rice and cotton. Research on rice is supported by the Rockefeller Foundation Programme on Rice Biotechnology. Cotton has been given a further boost through a recent US$10 million loan by the Asian Development Bank to MINFAL. Although transgenic plants have been developed by these two centres, work on field eval-

**Table 5.4.** National programmes on plant genetic engineering.

| Centre | Crop | Trait | Status | Commercial release |
|---|---|---|---|---|
| CEMB, Lahore | Cotton, rice, chickpea | *Bacillus thuringiensis* gene | Transgenic plants at testing stage | No |
| NIBRE, Faisalabad | Cotton | Antiviral gene | Transgenic plants at testing stage | No |
| | Rice | Salt-tolerant (GB uptake) | Transgenic plants at testing stage | No |

CEMB, Centre of Excellence in Molecular Biology.

uation is stalled due to the absence of biosafety rules. Further delay and uncertainty are expected due to the actual performance of genetically engineered crops in the field and their commercial development by public and private agencies. The present research efforts and their potential contribution are hard to assess. According to the most optimistic estimates, it will be at least 3–5 years before plants with desired traits can be produced and used in breeding programmes.

## International Biotechnology Initiatives

Bilateral and multilateral aid agencies, international organizations, national agricultural research institutions, universities, private foundations and commercial companies are all involved in the financing and/or execution of international biotechnology initiatives for developing countries (Brenner and Komen, 1994; Cohen, 1994). Since 1985, organizations covered by an Organization for Economic Cooperation and Development survey contributed an estimated US$260 million in grant funds to international biotechnology initiatives. These activities included: research programmes, advisory programmes, networks and specific projects. During this period, the total biotechnology component of World Bank loans and credits for national agricultural research projects in developing countries has been estimated at US$150 million.

Funding sources and expenditures relate to the crop and livestock biotechnology research programmes, for which comparative information is available. For these 25 programmes, total grant funding committed so far has totalled US$140 million. The overwhelming share of funding by foundations is provided by the Rockefeller Foundation. This foundation supports a number of smaller biotechnology research programmes, with the major share of its effort directed to the International Rice Biotechnology Programme. This programme involved an international network of researchers in universities and public research institutions in both industrial and developing countries, as well as at the International Rice Research Institute and the Centro Internacional de

Agricultural Tropical. Over US$80 million has been invested between 1985 and 2000.

Two countries contribute a large share of the total commitment to research programmes in agricultural biotechnology. These are France (through the plant and livestock programmes of the Centre de Coopération Internationale en Recherche Agronomique pour le Développement (CIRAD)) and the USA (through the US Agency for International Development (USAID)).

## Conclusion

The Pakistan Science Foundation (PSF) established a Task Force on Biotechnology in 1995, which resulted in many recommendations in various sectors, including agriculture and livestock. More recently, the Prime Minister's High Level Commission for Science and Technology, supported by the World Bank (1996/97), includes a standing committee on biotechnology.

In spite of the massive role of multinational companies in R & D of agricultural biotechnology, the national government still has major regulatory control over the testing, multiplication, distribution and safety of agricultural biotechnology products. Pakistan does not yet have a policy/legislation for intellectual property rights (IPRs) to cover patents involving biotechnology and biosafety codes for genetically modified organisms (GMOs).

Agricultural biotechnology must be considered as a supplement to existing crop improvement programmes. It is not a panacea, but a technology which, when backed up with other management strategies, is capable of delivering results. The following national projects, if taken up seriously, will definitely create an economic impact. These technologies are well known:

- Mass production of disease-free banana plants.
- Multiplication of virus-free potato.
- Multiplication of disease-free sugar-cane plants.
- Rapid multiplication of exotic clones of sugar cane.
- Rapid multiplication of female papaya, pineapple and other economically important fruits/flowers.
- Development of crops for biotic/abiotic stresses through genetic engineering.
- Initiation of molecular breeding work to speed up incorporation of desired traits through conventional breeding.

### Proposals for improvement

- Biotechnology/genetic engineering is specialized research requiring perishable enzymes, rare chemicals, clones, vectors and other resources. The perishable items have to be imported into Pakistan. The import policies, customs regulations and duty structure are lengthy, cumbersome and complicated. This makes it extremely difficult to obtain such materials, thus hindering developments in biotechnology R & D. In extreme cases, even donated material has perished during clearing processes, rendering it useless to the recipient. Heavy import duties (up to 55%) are placed even on donated research material. There is an urgent need to exempt all research material (equipment, consumables, material) from any customs, regulatory or sale duties/taxes.

- With the adoption of the Uruguay Round of the General Agreement on Tariffs and Trade (GATT), now the World Trade Organization (WTO), a new element was introduced: Trade-Related Aspects of Intellectual Property Rights (TRIPs). The TRIPs agreement sets the basis for the rapidly changing environment for the role of private firms, as it requires signatory states, including some 70 developing countries (Pakistan included), to provide added protection. Laws pertaining to plant breeding rights (PBRs) and IPRs must therefore be drafted and approved in Pakistan, one of the signatory countries commited to adopt within 10 years a full compendium of IPRs (Lele *et al.*, 2000).
- Biotechnology has been viewed by many political leaders, policy-makers and leading scientists in developing countries as a priority for nearly a decade. Nevertheless, the development of biosafety regulations has been slow. Pakistan still lacks regulatory mechanisms today. Effective biosafety regulations are critical for the safe and effective introduction of GMOs and transgenic crops and are indeed a prerequisite for the transfer of technology to developing countries. In Asia, China, India, the Philippines, Indonesia and Thailand have regulations in place, with Malaysia and Bangladesh having regulatory laws in the final drafting phase.

  In Pakistan, a draft has been prepared by NIBGE and submitted to the Ministry of Environment and Urban Affairs. There is an urgent need to develop biosafety capacity and regulatory mechanisms and to have these in place. International organizations, such as the UN Industrial Development Organization (UNIDO), the Food and Agriculture Organization (FAO), ICGEB, the Stockholm Environment Institute's Biotechnology Advisory Commission and the International Service for the Aquisition of Agri-biotech Applications (ISAAA), could be approached to help Pakistan in this regard.
- A team of senior scientists should form a think-tank team to formulate national strategies on science and technology. The PSF, Pakistan Academy of Sciences, Pakistan Council for Science and Technology and various science societies can play an active role in this initiative.

## References

Brenner, C. and Komen, J. (1994) *International Initiatives in Biotechnology for Developing Country Agriculture: Promises and Problems*. OECD Development Centre Technical Papers No. 100, OECD, Paris, France.

Cohen, J.I. (1994) *Biotechnology Priorities, Planning and Policies – A Framework for Decision Making*. ISNAR Research Report No. 6, International Service for National Agricultural Research, The Hague.

Khan, M.K. (1997) *Science and Technology for National Development: Fifty Years of Science and Technology in Pakistan*. PSF, Islamabad.

Khan, M.K. and Afzal, M. (1997) *Historical Perspective and Strategies for Technology Capacity Building*. PSF, Islamabad.

Lele, U., Lesser, W. and Horstkotte-Wesseler, G. (2000) *Intellectual Property Rights in Agriculture: The World Bank's Role in Assisting Borrower and Member Countries*. Rural Development Department, World Bank, Washington, DC.

# 6

# Philippines


## Reynaldo E. de la Cruz


| | |
|---|---|
| Area (km$^2$) | 300,000 |
| Cropland | 36,000 |
| Irrigated cropland | 15,800 |
| Permanent pasture | 12,000 |
| Population (1998 est.) | 79.3m |
| Population per km$^2$ | 264 |
| Annual population growth rate (1999 est.) | 2.04% |
| Life expectancy (men) | 63.8 yrs |
| (women) | 69.5 yrs |
| Adult literacy | 94.6% |
| GDP (1998 est.) | US$270 bn |

| | |
|---|---|
| GDP per head | US$3500 |
| Average annual growth in real GDP | |
| Annual inflation (1998) | 9.7% |
| Agriculture as % of GDP | 20% |
| Agriculture and food processing as % of manufacturing sector | 41% |
| Major export commodities: coconut, sugar cane, coffee, banana, pineapple, mango, fish | |
| Major subsistence commodities: rice, banana, coconut, maize | |

## Summary


Light industry and agriculture dominate the Philippine economy. There are well-developed industries in food processing, textile, clothing, wood–forest products and home appliances.

Key areas of science and technology identified by the government are biotechnology, materials science, information technology and energy research.

The Philippines was one of the first ASEAN countries to create institutes devoted to research and development in biotechnology. There are many highly trained scientists working in this area. The annual budget allocation for biotechnology is now about US$20 million.

In 1980 the Philippines created the National Institutes of Molecular Biology and Biotechnology (BIOTECH) at the University of the Philippines Los Baños (UPLB). Three other institutes were created in 1995, with the UPLB one continuing to provide leadership in agricultural, forestry, industrial and environmental biotechnology.

## Introduction

The combined area devoted to agriculture in the Philippines in 1977 was 10.3 million ha, with coconut being the most widely planted crop (4 million ha), followed by rice (3.5), maize (1.2), banana (0.2), pineapple (0.04) and others (1997 report of the Bureau of Agricultural Statistics). The production and value of some important agricultural crops are presented in Table 6.1. The country is a major producer of coconut, sugar cane, banana and pineapple. The export value of sugar cane and coffee has gone down considerably in recent years.

More than 70% of the population is directly or indirectly dependent on agriculture. Most of the land is owned by small-scale farmers. Increases in population have placed tremendous pressure on agricultural lands. Prime agricultural lands are being converted into resettlement areas and for industrial uses.

The Philippines started its biotechnology programmes in 1980 with the creation of the National Institutes of Molecular Biology and Biotechnology (BIOTECH) at the University of the Philippines (UP) at Los Baños (UPLB). In 1995, three other biotechnology institutes were established within the UP system. They are located in the UP Diliman campus for industrial biotechnology, UP Manila for human health biotechnology and UP Visayas for marine biotechnology.

**Table 6.1.** Production and value of some important agricultural crops (from Bureau of Agricultural Statistics Report, 1997).

| Agricultural crops | Production in million metric tonnes | Value in US$ millions |
|---|---|---|
| Rice and maize | 26.9 | 33 |
| Coconut | 12.0 | 8 |
| Sugar cane | 3.4 | 1 |
| Banana | 21.6 | 4 |
| Pineapple | 1.6 | 3 |
| Coffee | 0.1 | 1 |
| Others | 2.4 | 60 |
| Total | 68.0 | 110 |

The UPLB institute continues to provide leadership in agricultural, forestry, industrial and environmental biotechnology. Other research institutes at UPLB are also doing biotechnology research. Among these are the Institute of Plant Breeding, Institute of Biological Sciences, Institute of Animal Sciences, Institute of Food Science and Technology and College of Forestry and Natural Resources. Outside UPLB, other research institutes and centres, such as the Philippine Rice Research Institute, Philippine Coconut Authority, Cotton Research and Development Administration, Bureau of Plant Industry, Bureau of Animal Industry and Industrial Technology and Development Institute, are also involved in biotechnology R & D (see also Novenario-Enriquez, 2000).

The type of research undertaken in the Philippines from 1980 to 1999 was mainly conventional biotechnology, with the exception of a small amount of work on molecular markers and the development of genetically improved organisms (GIOs) with useful traits (Table 6.2).

In 1998, five high-level biotechnology research projects were funded by government:

1. Transgenic banana resistant to banana bunchy-top virus and papaya resistant to papaya ringspot virus.
2. Delayed ripening of papaya and mango.
3. *Bacillus thuringiensis*-maize.
4. Marker-assisted breeding in coconut.
5. Coconut with high lauric acid content.

About 80% of the total annual budget for biotechnology R & D comes from the government, 15% from international development agencies and 5% from the private sector. The last is expected to provide more funding in future as they see the potential of biotechnology in agriculture.

In 1997, the Agriculture Fisheries Modernization Act (AFMA) became

**Table 6.2.** Biotechnology projects funded from 1980 to 1999 (from survey conducted by UPLB BIOTECH, 1999).

| Type of R & D | No. projects | Per cent of total |
|---|---|---|
| Biocontrol | 55 | 20.9 |
| Soil amendments | 44 | 16.7 |
| Food/beverage | 43 | 16.3 |
| Tissue culture | 52 | 19.7 |
| Feed component | 20 | 7.6 |
| Enzymes | 16 | 6.0 |
| Diagnostics | 7 | 2.6 |
| Farm waste utilization | 4 | 1.5 |
| Vaccines | 3 | 1.1 |
| Animal reproduction | 3 | 1.1 |
| Molecular markers | 12 | 4.6 |
| GIOs | 7 | 2.7 |
| Total | 266 | 100 |

law. The main objective of AFMA is to modernize agriculture, including infrastructure, facilities and R & D. AFMA recognized biotechnology as a major strategy to increase agricultural productivity. The law states that AFMA will provide a budget of 4% of the total R & D budget per year for biotechnology during the next 7 years. This allocation provides an annual budget for biotechnology of almost US$20 million. Before AFMA, the annual budget averaged less than US$1 million.

AFMA operates through national research, development and extension (RDE) network systems of 13 commodities and five disciplines. The 13 commodity networks are rice, maize, root crops, coconut, plantation crops, fibre crops, vegetables/spices, ornamentals, fruit/nuts, capture fisheries/aquaculture, livestock and poultry, and legumes. All of these commodities include biotechnology in their RDE agenda. The five discipline-orientated RDE networks are fishery postharvest and marketing, soil and water resources, agricultural and fisheries engineering, postharvest, food and nutrition, social science and policy, and biotechnology. Biotechnology focuses on upstream basic research, which includes work in molecular biology. The commodity networks focus on downstream (application) research.

The main goal of biotechnology R & D under AFMA is to harness the potential of this cutting-edge technology to increase productivity of all the commodities in the agriculture and fishery sectors. Biotechnology will therefore play a major role in the selection and breeding of new varieties of plants and animals. It will also provide the inputs required such as biofertilizers and biocontrol of harmful pests and diseases. Biotechnology will also be used to produce genetically improved crops with resistance to harmful pests and diseases, for accurate diagnosis and control of diseases in plants and animals, for bioremediation of the environment and for bioprospecting. The benefits derived are intended to reach the small farmers and fishermen.

Adequate laboratory facilities and equipment for upstream biotechnological research exist at a number of institutions in the Philippines, including UPLB BIOTECH and UP Diliman, the Institute of Biological Sciences, the Institute of Plant Breeding and the Philippine Rice Research Institute. There is a need, however, to upgrade most of the laboratories in the country. The Philippines does not have adequate human resources required for biotechnology R & D. As of 1999, there were about 250 scientists qualified to do high-level biotechnology R & D. Most of the researchers are affiliated with universities, particularly UPLB.

## Opportunities

Although the Philippines is lagging behind the industrial countries and their ASEAN neighbours in terms of R & D in biotechnology, many windows of opportunity are open. Biotechnology provides the opportunity for researchers to improve plant growth, development and yield by providing for the basic needs of the plant, such as biofertilizers and biocontrol agents.

The country recognizes the tremendous potential of improved crop plants containing genes that provide pesticidal properties, resistance to herbicides, tolerance to pests, diseases and stress (salt, heavy metals and drought), or combinations of these. Such improved plants are expected to reduce production costs. Once the issues of biosafety regulations and intellectual property have been settled, the country will be able to use such new plant technologies, which are now limited to only a few countries.

Marker technologies may help speed up the selection and production of more effective hybrids. Most breeding work in the Philippines is now using this technology, particularly in rice, maize and coconut.

Tremendous opportunities are available for livestock biotechnology, including the production of vaccines for foot-and-mouth disease and haemorrhagic septicaemia, for diagnostics and for *in vitro* fertilization.

Opportunities are available to use microorganisms for biofertilizers, biopesticides and bioremediation of the environment.

The Philippines is blessed with rich genetic resources waiting to be tapped for food, fibre, enzymes and drugs. New beneficial genes are expected to be discovered in the highly diverse species of plants, animals, microorganisms and marine organisms. The challenge is to save and use judiciously the rich biodiversity of the country.

This biodiversity offers many opportunities in the search for novel genes and gene products. The Philippines has in place a law governing access to genetic resources by foreign and local bioprospectors. This law is designed to protect the bioresource and the bioprospectors.

Because of the importance given to R & D in biotechnology under AFMA, the introduction of foreign technologies, including genes that offer unique advantages, may have great potential for the country. For example, the sugar industry had been declining because of competition with high-fructose syrup and other sugar substitutes. There are opportunities to use sugar cane, a highly efficient plant, to produce high-value products, such as oral vaccines, biodegradable plastics and other products.

Collaboration between Philippine and overseas researchers is one opportunity that is now well in place. Many researchers actively collaborate with researchers from Australia, Canada, the USA, Japan, South Korea and countries of the European Union.

## Challenges

Although the country recognizes the tremendous potential that can be achieved from biotechnology, several challenges need to be met before the goals set can be achieved.

Yields of crops and livestock have been declining, while demands are increasing because of the rapid increase in population. Conversion of prime agricultural lands into other uses has placed tremendous pressure on the agricultural sector to increase productivity per unit area. Productivity has been affected by poor soil fertility, the incidence of pests and diseases and abiotic stresses, such as drought and climatic factors, especially typhoons. The challenge is to use biotechnology to increase productivity and yield on the farms using minimal inputs.

With impending trade liberalization, the country expects to receive cheap agricultural products from other countries, thus widening its balance of trade. In 1998, the value of Philippine exports was estimated at US$28 billion, while imports were valued at US$29 billion. The challenge is to use biotechnology to produce local products that are highly competitive with those from foreign sources, thereby promoting exports of quality products while reducing imports.

The Philippines is sensitive to the issue of biosafety, having one of the strictest biosafety guidelines in the world to undertake R & D and for field testing. The challenge is to improve and better implement the current biosafety guidelines, taking advantage of knowledge generated worldwide. Protocols are needed to assess the risk of GIOs and to manage any identified risk factors. The Philippines must develop capabilities to undertake risk assessments and management, based on scientific evidence.

The commercial release of new products must be regulated. At present, all regulatory bodies, such as the Bureau of Plant Industry, Bureau of Animal Industry, Fertilizer and Pesticide Administration, Bureau of Food and Drugs Administration and Environment and Management Bureau, lack policies and guidelines to regulate the commercial release of new genetically improved products. In addition, the institutional support system, such as laboratories and infrastructure, is not in place. The challenge is to create guidelines to regulate commercialization of GIOs, the establishment of support laboratories and infrastructure and the training of people for these regulatory bodies.

Products of research will not create any measurable impact unless they are transferred to end-users and/or commercialized. The challenge is to transfer products to users, particularly to small farmers and fishermen. This requires the proper packaging of the product to attract private investors for eventual commercialization.

Transgenic crops and other genetically improved products may become trade-related issues in the future because of trade liberatization. It is expected that new genetically improved crops will be imported into the Philippines. The challenge is to create public awareness of the benefits and risks of any new product and to assist in the availability of new and beneficial technologies to consumers.

Because the processes, products and genetic materials used in biotechnology R & D have proprietary considerations, issues of intellectual property protection by patents and plant variety protection will arise. The present

Intellectual Property Code of the Philippines allows the patenting of microorganisms, but not plants and animals. Plant varieties will be protected by a *sui generis* mechanism after both houses of Congress pass the plant variety protection bill. The challenge is for the country to strengthen its intellectual property laws to provide protection for researchers, discoverers and investors.

## Constraints

Although the R & D opportunities are evident, there are some constraints that need to be addressed.

Development of the local biotechnology industry has been hampered because of the inability of researchers to access state-of-the-art technologies. Researchers are therefore repeating work done elsewhere rather than being able to adopt current technologies.

Some non-governmental organizations (NGOs) and individuals in academe and government services do not support biotechnology. These groups are well organized and well funded and are highly successful in promoting anti-biotechnology sentiments in the country. They are also instrumental in persuading legislators to enact resolutions imposing moratoria on the research and commercialization of GIOs. Although they focus on genetically improved products produced and brought into the country by multinational companies, they also affect the R & D of local researchers.

The present set of biosafety guidelines is one of the strictest in the world. The guidelines were originally patterned after those first used in the USA, Australia and Japan during the early 1980s. Since then, all these countries have relaxed most of their guidelines as a result of new technical data and familiarity in dealing with new products. However, the Philippines did not relax their guidelines.

New genetically improved products cannot be commercialized in the country because the regulatory bodies cannot issue the required permits or licences. The regulations allow only limited field trials of GIOs. The regulatory bodies lack the proper guidelines and institutional support to regulate the new products. This is a major constraint.

## International Cooperation

The international agricultural research centres (IARCs) can play a larger role in helping national centres to develop their R & D capabilities in biotechnology. Some activities that IARCs can undertake include the following:

- Many IARCs hold extensive collections of germ-plasm, the starting-point for selection, breeding and genetic manipulation. The IARCs are in a position to share or exchange this germ-plasm with local researchers or institutes.

- IARCs should encourage more joint collaborative research with local institutes and share their financial and human resources and infrastructure with less well-endowed local research institutes. IARCs are also in a position to assist human resource development through training, workshops and scholarships.
- IARCs should help countries develop their biosafety protocols and competence in risk assessment and management of biotechnology products. IARCs may also be able to assist countries in developing regulatory mechanisms and institutional capabilities for the commercialization of biotechnology products.
- IARCs could be more proactive in promoting popular awareness and acceptance of the products of modern biotechnology.

## Conclusions

Researchers, policy-makers, people from industry and the IARCs must address the challenges, opportunities and constraints that face R & D in biotechnology. All countries share these same challenges, opportunities and constraints, although at different levels.

The challenges, opportunities and constraints can be addressed by international agencies at the international level and by national R & D centres at a country level, with harmonized activities at international, regional and country levels.

For developing countries, the small farmers and fisherfolk should be the main beneficiaries of biotechnology R & D. Biotechnology will only prosper if the private sector actively participates in both the R & D and commercialization stages.

## Reference

Novenario-Enriquez, V.G. (2000) Philippines. In: Tzotzos, G.T. and Skryabin, K.G. (eds) *Biotechnology in the Developing World and Countries in Economic Transition.* CAB International, Wallingford, UK, pp. 147–157.

# Thailand

7

**Morakot Tanticharoen**

| | |
|---|---|
| Area (km²) | 514,000 |
| Cropland | 34% |
| Irrigated cropland (km²) | 44,000 |
| Permanent pasture | 2% |
| Population (1999 est.) | 60.6m |
| Population per km² | 117 |
| Annual pop. growth rate (1999 est.) | 0.9% |
| Life expectancy (men) | 65 yrs |
| (women) | 73 yrs |
| Adult literacy | 93% |
| GDP (1998 est.) | US$369 bn |
| GDP per head | US$6100 |
| Agriculture as % of GDP | 16% |
| Value of agricultural exports (1997) | US$7.2 bn |
| Agricultural products as % of total exports (1998) | 23% |

Major export commodities:
rice, rubber, prawn, cassava

Major subsistence commodities:
rice, cassava, maize, coconut, soybean

## Summary

Before the economic crisis of 1997, Thailand was named one of the Asian newly industrialized countries (NICs).

Efforts to revive the economy are currently under way in both the government and the private sector. The linkage between the status of science and technology and the economic status of a country has long been recognized. It is crucial for Thailand to increase the technological capability of the country, to make efficient use of its resources and to reduce the cost of production, thereby increasing economic growth and competitiveness.

Despite the country's industrialization, agriculture has remained a significant part of the economy. Thailand has been moving towards industrial-based agriculture and has focused on the development of postharvest and processing technologies that are the major problems for industry. Biotechnology has become the country's priority for research and development and for the benefit of the private sector, as well as rural development.

## Introduction

The agriculture sector expanded by about 2.8 % in 1998, although most of the economic sectors registered negative growth rates. Thailand's Ministry of Agriculture estimated that farmers would earn 650 billion bahts (US$16.2 billion) for the year, of which 74% would come from the major products listed in Table 7.1.

The government promotion to develop agribusinesses since 1976 has greatly contributed to the expansion of agroprocessing. Combined export earnings from agriculture accounted for 23% of total earnings (Department of Business Economics). Thailand's top ten food exports in 1998 are given in Table 7.2.

Recent exports have been hit by tough price competition from lower-wage Asian countries. The result showed that Thailand could not depend solely on its weaker currency to boost exports. To remain competitive, Thailand will have to focus more on the country's development and be more innovative and

**Table 7.1.** Production of key agricultural products and earnings in 1998/99.

| | Earnings* (US$ billion) 1998/99* | Production† (million metric tonnes) 1998/99* |
|---|---|---|
| Rice | 3.27 | 21.50 |
| Black tiger prawns | 1.42 | 0.20 |
| Rubber | 1.18 | 2.31 |
| Pigs | 1.23 | No data |
| Sugar cane | 0.60 | 42.60 |
| Cassava | 0.45 | 16.37 |
| Chicken | 0.70 | 0.84 |
| Maize | 0.50 | 4.99 |
| Chicken eggs | 0.38 | No data |
| Oil-palm | 0.02 | 2.67 |
| Soybean | 0.009 | 0.37 |

*Commerce Ministry.
†Ministries of Commerce and Agriculture.

**Table 7.2.** Thailand's top ten food exports in 1998.

| | Export value (US$ billion) |
|---|---|
| Rice | 2.17 |
| Canned fish | 1.69 |
| Fresh chilled/frozen shrimps, prawns and lobsters | 1.45 |
| Sugar | 0.66 |
| Tapioca products | 0.57 |
| Chilled/frozen poultry cuts | 0.41 |
| Prepared/preserved fruits in airtight containers | 0.38 |
| Fresh chilled/frozen cuttlefish, squid and octopus | 0.29 |
| Prepared/processed foods for animal feeds | 0.25 |
| Processed poultry | 0.22 |

creative in R & D. Improving crop yield and protecting agricultural crops from diseases and pests, improving postharvest handling and diversifying products are all priorities for Thailand. There is a need to improve productivity of Thai crops, while retaining their unique qualities (for example, the fragrant Thai rice Khao Dawk Mali). Rice yield in Thailand averages only 2.4 t $ha^{-1}$, compared with 6.3, 6.0, 4.3 and 3.6 t $ha^{-1}$ in the USA, China, Indonesia and Vietnam, respectively.

Thai sugar-cane yields are only 48.8 t $ha^{-1}$, compared with 93.8 in Brazil. The country's 46 sugar-mills, meanwhile, have the capacity to process more than double the amount of cane they now receive. Another problem with Thai cane is the sweetness. The international grading system has given a rating of 11 ccs (commercial cane sugar) for Thai sugar compared with 13–14 for other countries. The Office of the Cane and Sugar Board's main activity at the moment is to develop better breeds, with the goal of increasing the sweetness grade of Thai cane to 15 within 5 years. The new strain should also be resistant to drought, salty soil and diseases.

A master plan for Thailand's agricultural development was approved by the government in early 1998 to make exports more competitive. The objectives are supported by a master plan for industrial restructuring, approved in April 1998. Thirteen industries will be promoted to make Thailand an important export centre in Asia within 2 years. Three industries using agricultural products (food and animal feed, rubber and rubber products and wooden products, including furniture) are included in 13 industries. Key agricultural projects are as follows:

- The establishment of integrated agricultural zones for exports.
- R & D to raise production and cut costs by using new technology with emphasis on biotechnology. Rice, livestock, rubber, durian, longan and orchids are priority commodities.
- Bringing product quality and processing up to international requirements.

A centre to control quality from the raw-material stage to the finished product will be established.

- Restructuring the Agriculture Ministry to modernize its management and services.
- Encouraging farmers to use less chemical fertilizer while promoting natural alternatives and organic production.
- Improving management of land use and ownership, natural resources, irrigation and coastal areas.
- The establishment of weather warning systems in high-risk areas.
- Improving farm methods and technology.

The Agriculture Ministry outlined five strategic plans for 1999, with a budget of about US$1 billion:

- Increase competitiveness of farm products for export and import substitution (US$305 million) and to promote self-sufficient farm projects (US$24 million).
- Management of natural resources and the environment (US$372 million).
- Development of an agricultural institute (US$225 million) to encourage community-based production.
- Plans initiated by His Majesty the King (US$78 million).
- Preparation for the 21st century (US$4 million).

Apart from the government's annual budget, the ministry has obtained US$600 million, mainly from the Asian Development Bank (ADB), to improve the agricultural economy through a series of short- and long-term programmes.

## Opportunities

### BIOTEC

The National Centre for Genetic Engineering and Biotechnology (BIOTEC) was established under the Ministry for Science, Technology and Energy in September 1983. In 1991, Thailand established the National Science and Technology Development Agency (NSTDA) and BIOTEC became one of the NSTDA centres, operating autonomously outside the normal framework of civil service and state enterprises. This enabled it to operate more effectively to support and transfer technology for the development of industry, agriculture, natural resources, the environment and the socio-economy (Sriwatanapongse *et al.*, 2000).

BIOTEC policy provides the resources for the country to develop the critical mass of researchers necessary to achieve Thailand's national R & D requirements in biotechnology. This is achieved through R & D projects, the facilitation of transfer of advanced technologies from overseas, human resource development at all levels, institution building, information services and the development of public understanding of the benefits of biotechnology.

BIOTEC is both a granting and implementing agency. It allocates approximately 70% of its R & D budget to several universities and research institutes in Thailand and 30% for in-house research projects. The facilities of national and specialized laboratories are made available for in-house research programmes as well as for visiting researchers. The construction of a Science and Technology Park will be completed in 2001 and will house BIOTEC's main laboratories.

Several research programmes have been undertaken by a BIOTEC-appointed committee of recognized experts in the field. The major biotechnology programmes and activities are described below.

### *Shrimp*

Basic knowledge about the major cultivated shrimp species has lagged behind technical innovations that have led to successful intensification of culture and to ever-increasing world production. Sustaining high production levels will require innovation to minimize adverse environmental impacts. Biotechnology will play a central role in helping us to know more about shrimp and thereby improve rearing practices. BIOTEC's support will focus on issues dealing with shrimp diseases and with improvement of the seed supply. The disease work has so far emphasized the characterization, diagnosis and control of serious shrimp pathogens, particularly yellow-head disease (YHD) and white-spot syndrome (WSS) disease. Luminescent bacterial infections have contributed to the declining production to a lesser degree. These diseases become progressively more serious threats to the industry as it has grown and intensified. Indeed, the work on YHD virus and WSS virus (WSSV) supported by BIOTEC has been instrumental in substantially reducing losses in Thailand. The losses to YHD (probably exceeding US$40 million in 1995) and those to WSS (probably exceeding US$500 million in 1996) could have been much more serious without the basic knowledge and the DNA diagnostic probes made available to the industry by Thai researchers. Checking for subclinical WSSV infections by polymerase chain reaction (PCR) has been a common practice in Thailand, to help farmers in screening out WSSV +ve postlarvae (PL) before stocking (Flegel, 1997).

The Shrimp Biotechnology Service Laboratory was established in July 1999 at BIOTEC to summarize the reference PCR methods for viral disease detection in Thai shrimp farming. Laboratory objectives are to serve as the reference laboratory for major shrimp pathogen diagnosis based on molecular techniques, to conduct research and to provide assistance for the molecular detection of various shrimp viruses.

It has been reported that WSSV can be vertically transmitted and widespread among wild brood-stock. In addition to the disease problem, a decline in the growth rate of shrimp produced from currently available wild brood-stock has also been observed. Production of specific pathogen-free animals and the development of specific pathogen-resistant (SPR) strains are now being carried out in the USA, Venezuela and French Polynesia with *Penaeus stylirostris* and *Penaeus vannamei*. This could be considered a breakthrough, since production of *P. vannamei* more than doubled during 1992–1994. Currently the most important programme involves the domestication and genetic improvement of *Penaeus monodon* stocks (Withyachumnarnkul *et al.*, 1998). The project will lead to the development of SPR stocks and improved growth performance through selective breeding. The first domesticated stocks from this programme were to be ready for pond production tests in 1999. BIOTEC is also supporting advanced studies on DNA characterization and DNA tagging of the shrimp stocks. These studies are providing the tools that will be important for rapid genetic improvement strategies.

BIOTEC is dedicated to the principle that the players in the shrimp industry should take an active role in the R & D effort for their industry, in both planning and finance. BIOTEC took an active part in promoting the formation in 1996 of an industry consortium (the Shrimp Culture Research and Development Company) dedicated to solving problems common to the shrimp aquaculture industry as a whole. This consortium serves the industry directly and also serves as a bridge to other public and private institutions involved in relevant research, not only in Thailand, but throughout the world.

### *Cassava and starch*

About 70% of the 16 million t of cassava root produced in 1998 was used in the production of pellets and chips, and the remaining 30% was mainly used to produce flour and starch. A production shortage in 1997/98 prompted the Thai Tapioca Development Institute (TTDI) and Kasetsart University to develop a new strain with a higher yield. Kasetsart 50, a new strain, has an average yield of 26.4 t of roots $ha^{-1}$ and a starch content of 26.7%, compared with 13.75 t $ha^{-1}$ and 18% starch content of the best strain available.

The tapioca starch industry is one of the largest in Thailand. In 1998, tapioca starch was worth about US$120 million. About 40% of starch was used domestically for the production of modified starch, sweetener and monosodium glutamate. Most of the remaining 60% was exported. Efficient production, low production costs and the development of value-added products are vital to the starch industry and the farming sector (total of 1.3 million ha planted in cassava). The programme on starch and cassava products was established to provide support and funding for R & D. The programme is funded jointly by BIOTEC and TTDI to carry out R & D in three core activities. The short-term project aims to improve the processing efficiency of starch production, in particular to minimize water and energy consumption. This will reduce water use and costs and also reduce waste-water treatment. Waste-water discharge varies from 13 to 50 $m^3$ $t^{-1}$ of starch produced, with an average of 20 $m^3$. A benchmark on water use is a priority for the Thai starch industry.

Biotechnology can play an important role in waste utilization. Solid waste (after starch extraction) still contains 50% starch (dry weight) and has been utilized as animal feed. Tapioca, however, is not suitable for the production of feed requiring high protein content. Attempts have been made for protein enrichment using various microorganisms, such as *Aspergillus* and *Rhizopus*. Nevertheless, the economic feasibility is still in doubt and further technological development is needed. In contrast, turning waste water into energy through high-rate anaerobic digestion is promising. Though the technology is proven, an adaptation to such high-strength waste water and low buffering capacity is required to ensure stability of the system. In comparison with the upflow anaerobic sludge blanket (UASB) technology, the fixed bed is easier to control and operate. R & D, however, is focused on increasing loading efficiency. Based on calculations, methane generated from anaerobic treatment of starch waste water from 60 factories would be approximately 630 million $m^3$ annually. This could be substituted for fuel oil used in drying, saving energy costs of about US$4 million annually. There is also the environmental cost of large land areas required for conventional pond systems. In addition to native starch, production of modified starch is increasing, leaving an excessive amount of sulphate in waste water. This may interfere with the anaerobic digestion intended for energy production. A number of papers have been published recently on the interactions between the sulphate-reducing bacteria (SRB) and the methanogenic bacteria (MGB). Molecular diagnosis has been developed and applied for the mixed cultured system. A better understanding of these anaerobic microbes could lead to the biological removal of sulphate, which is the main problem of various industries.

The European Union (EU) has set a quota for exported tapioca pellets. Product diversification is part of the second core research activity. As a result, production of biodegradable plastic from cassava starch is being investigated. Increasing use of cassava as a raw material for fermentation industries, such as amino acids and organic acids, must proceed, expanding the development of value-added products. To reduce costs of production, however, research is orientated towards the production of good-quality cassava chips as a starting material to replace the starch.

Finally, basic research on cassava starch structure and properties will add to our knowledge and help increase the use of cassava starch. The Cassava and Starch Technology Unit, a specialized BIOTEC laboratory established in 1995 at Kasetsart University, has been engaged in studying the physicochemical properties of cassava. The unit is well equipped and provides regular service and training on instrument analysis of starch properties for the private sector and government agencies.

### *Rice*

Rice yields in Thailand are low. One of the major constraints in cultivation is blast disease, especially in high-quality rice cultivars, such as the aromatic Khao Dawk Mali. In northern Thailand, about 200,000 ha of rice were affected by blast in 1993, causing serious economic loss and resulting in government intervention of about US$10 million to assist disease-struck farms.

Another US$1.2 million was spent on fungicides (Disthaporn, 1994). Attempts have been made to breed higher resistance levels to blast in Thai rice. Limiting factors, however, are lack of insight and information on resistance genes and the complex structure of the pathogen populations. Genetic analysis provides an efficient tool to identify useful resistance genes in the host while analysing the race composition of the pathogen population. Recent research activities applying molecular genetic methods (DNA fingerprinting of a blast isolate collection at Ubon Ratchathani Rice Research Station, mapping of host resistance genes by the DNA Fingerprinting Unit at Kamphaengsaen campus of Kasetsart University) are providing baseline data on the interaction between rice and blast. The project is working on three closely related areas as follows:

- Establishment of a suitable differential cultivar series; identification of resistance genes conferring complete and partial resistance to blast disease in rice.
- Pathotype and molecular genetic characterization of the blast pathogen population in Thailand. So far, more than 500 monospore isolates have been deposited with the BIOTEC specialized culture collection.
- The special case of fertile isolates: the potential of using Thai isolates of *Magnaporthe grisea* for the development of a molecular diagnostic tool for pathogen race analysis. The degree of fertility can be assessed from the timing and number of perithecia that develop. BIOTEC has the capacity to test the mating type of about 80 isolates per month.

This project is a nationwide, network-type collaboration combining molecular genetics and classical approaches to help scientists breed rice cultivars with improved blast resistance.

BIOTEC provided US$1.5 million in 1999 to fund the Rice Genome Project Thailand. On behalf of Thailand, BIOTEC has joined an International Collaboration for Sequencing the Rice Genome (ICSRG) by sequencing 1 Mb annually of chromosome 9 for the next 5 years. BIOTEC is expected to provide about US$3.7 million to cover this work. Chromosome 9 was selected based on the previous extensive work on the fine genetic and physical maps surrounding the submergence-tolerance quantitative trait locus (QTL), the prospect of gene richness and the small chromosome size. Joining ICSRG will allow Thai scientists to directly access the rest of the genome sequence made available by the other collaborating members. Gene discovery from wild rice germ-plasm will be undertaken in parallel to efficiently use the genome sequence data. The project will bring Thailand into the international scientific arena, incorporate state-of-the-art technology and improve Thailand's competitive edge in the international rice market.

### *Dairy cows*

In 1997, Thai milk consumption was 12 l person$^{-1}$ year$^{-1}$. Milk production is still insufficient to meet local demand and Thailand has to import more than

50% (worth US$305 million) of the dairy products consumed in the country. To meet the national demand, we need an additional 130,000 cows.

Reproductive efficiency is a primary determinant of dairy herd production profitability. Milk yield (10 kg $day^{-1}$) is still far below the average (30 kg $day^{-1}$) of most developing countries. It is therefore important to promote an increase in dairy production through science and technology. The major programmes are breeding and feeding. The lack of proper management is another major contributing factor to an underproductive dairy industry.

Traditional breeding practices in Thailand have been too slow to meet national requirements and importing pregnant heifers and/or young quality-bred calves from abroad is too costly. Cutting-edge technologies, such as embryo transfer, *in vitro* fertilization, embryo sexing and semen sexing, have been studied by Thai scientists for more than 10 years. Nevertheless, the technologies have not yet been adopted. Technology transfer and training of Thai researchers at the leading laboratories/companies are now under discussion. The goal is to increase production of high-quality heifer calves at the most economical cost.

### *Gene engineering*

By the mid-1970s, with biotechnology centred in genetic engineering and molecular biology, Thailand was ready to adopt the new tools and apply them to various practical problems, first in the biomedical field and later in agriculture and other areas. A few specific examples will be given here to highlight the application of molecular biology and genetic engineering to agricultural development. Efforts in agricultural biotechnology and genetic engineering have been focused on three main areas: plant transformation, DNA fingerprinting and molecular diagnosis of plant and animal diseases. The first area should lead to the production of transgenic plants with superior properties, including resistance to diseases, insect pests and abiotic stresses.

The Plant Genetic Engineering Unit (PGEU), the specialized laboratory of BIOTEC at Kasetsart University, Kamphaengsaen campus, was established in 1985 to carry out work on plant biotechnology and genetic engineering. A transgenic tomato plant carrying the coat-protein gene of tomato yellow-leaf-curl virus was first developed to control this serious virus disease of tomato (Attathom *et al.*, 1990). The same approach was taken to develop transgenic papaya and pepper for resistance to papaya ringspot virus and chilli vein-banding mottle virus, respectively (Chaopongpang *et al.*, 1996; Phaosang *et al.*, 1996). Sri Somrong 60, a Thai cotton variety, was successfully transformed with cryIA[b] gene expressing a toxin from *Bacillus thuringiensis* (Bt). Development of transgenic rice varieties has been supported by the Rice Biotechnology Programme launched by BIOTEC and the Rockefeller Foundation. An example is the transformation of Khao Dawk Mali 105, an aromatic Thai rice with $\Delta^1$ -pyrroline-5-carboxylate synthetase (P5CS) for salt and drought tolerance. Most transgenic plants are now being tested under greenhouse conditions in accordance with the Biosafety Guidelines (Attathom and Sriwatanapongse, 1994; Attathom *et al.*, 1996). Field testing of transgenic plants developed in Thailand will begin in 2001.

### *DNA fingerprinting*

Each living creature has a unique DNA sequence. Using DNA fingerprinting and PCR, scientists can identify organisms and genes. Important genes can be located (genetic maps). Moreover, the availability of DNA probes and specific sequences has made it possible to develop appropriate molecular methods for the diagnosis of plant and animal diseases. Molecular mapping of genes in rice involving flooding tolerance, rice blast, aroma, cooking quality and fertility restoration were accomplished using three mapping populations. A back-cross breeding programme for the improvement of Jasmine rice was initiated. In the first stage, resistance to bacterial leaf blight, flooding tolerance, resistance to brown planthopper/gall midge and photoperiod insensitivity were the main areas of focus. Restriction fragment length polymorphism (RFLP)-based markers were an important limiting factor for high throughput and cost-effectiveness. The PCR-based marker for Xa21 is the most reliable for marker-assisted back-crossing in rice.

Tomato production in the tropics and subtropics faces serious constraints due to bacterial wilt (BW), a disease caused by the bacterial pathogen recently reclassified as *Ralstonia solanacearum* (previously *Pseudomonas solanacearum*). In Thailand, an endemic outbreak of BW in tomato, potato, pepper, ginger and groundnut occurs each year, causing a yield loss of approximately 50–90% depending on growing conditions. BW-resistant varieties cannot easily be developed, due to the nature of the (quantitatively inherited) resistance, which involves several genes. Marker-assisted selection (MAS), a breeding method of selecting individuals based on markers linked to target genes, in addition to phenotypic measurement, is essential and useful only for enhanced resistance to diseases. At this time, three putative QTLs corresponding to BW resistance have been found, using amplified fragment length polymorphism (AFLP) markers. Once markers closely linked to BW-related QTLs are well established, they can be used for marker-assisted breeding for enhanced resistance to bacterial wilt in tomato. A tomato consortium has been set up to extend public–private collaboration.

BIOTEC has set up the DNA Fingerprinting Service Unit at Kasetsart University. The unit has provided services for public and private concerns for more than 2 years. The main services are DNA fingerprinting and DNA diagnosis.

### *Biocontrol*

In 1996, Thailand imported 38,000 t of chemicals, mainly insecticides and herbicides. The global trend of going organic is an opportunity for Thai farmers to supply fresh organic produce, especially fruit and vegetables, to the world. Over the past decade, the developmental work on biocontrol in Thailand has continued to receive active support from BIOTEC and the Thailand

Research Fund. Two companies are now producing commercially grown *Trichoderma* to control *Sclerotium rolfsii* Sacc. and *Chaetomium* to control soil fungi, such as *Phytophthora* (Yuthavong, 1999). BIOTEC and the Department of Agriculture have set up a pilot-scale production facility to produce nuclear polyhedrosis virus (NPV), Bt and *Bacillus sphericus*. NPV is widely used to control *Spodoptera* moth in grapes. Bt produced locally has gained popularity over the last few years. The capacities of pilot plants at Mahidol University and King Mongkut's University of Technology (Thonburi) are taken up with Bt production. Commercial production may begin soon. A project at Mahidol University to transfer the chitinase gene into *B. thuringiensis* subsp. *israelensis* has received support from BIOTEC.

## Biosafety

Biosafety issues are being debated in Thailand. The National Biosafety Committee (NBC) was established in January 1993 under BIOTEC. The NBC has introduced two biosafety guidelines: one for laboratory work and the other for fieldwork and the release of genetically improved organisms (GIOs) into the environment. The establishment of institutional biosafety committees (IBCs) at various public institutes and private companies was also strongly recommended and, in many cases, these recommendations have been implemented.

The importation of prohibited materials under Plant Quarantine Law BE 2507 implemented by the Department of Agriculture controls to a certain degree the use of GIOs. Ministry regulation II (1994 [A.D.]) identifies certain prohibited transgenic plants. Permission from the Ministry of Agriculture is required to perform field testing of transgenic plants brought into Thailand. The following have received permission to perform the test: the Flavr Savr tomato produced by Calgene for the production of seeds (1994); a field trial of Monsanto Bt cotton was carried out under restricted containment in a netted house in 1996; in 1997, a Bt maize field trial was approved to be carried out by Novartis at their experiment station in a netted screen house.

People seem to pay more attention to the introduction of GIOs into the country by the multinational companies than to considerations of technological information. An issue never discussed or debated, in particular at the political level, is whether or not Thailand should be more aggressive on the development of transgenic organisms. Thailand is rich in biodiversity and several genes resistant to biotic and abiotic stresses embedded in wild plants and other bioresources need to be discovered and utilized. This illustrates the potential benefits of biotechnology and genetic engineering. In the 1980s, when genetic engineering and biotechnology first made their impact felt, genetic engineering capability was present in only two or three institutions in Thailand (Yuthavong, 1987). Ten institutions now have genetic engineering capability. Nevertheless, the most important challenge for the future of GIOs is not technical in nature, but the attitude of the public towards the technology. These issues need to be studied and debated among the scientists, the public and the policy-makers, and an optimal policy developed. BIOTEC realizes that genetic engineering depends critically on public support, so the Centre has emphasized public education, with information programmes on GIOs being introduced to the public and to industry.

## Conclusions

Although Thailand is a leading exporter of food products, it also imports food commodities that are not available or that cannot be adequately supplied through local production. Among Thailand's top ten food imports in 1998 are fresh and frozen tuna used for canning and vegetable materials for animal feed preparation. Exports of frozen and processed chicken are expected to remain at 1998 levels of 140,000 t for the next 2 years. Maize, soybean meal and fish-meal are key ingredients for feed industries. Maize production for the 1998/99 crop year will be approximately 4.9 million t, whereas local demand, mainly from animal feed factories, was expected to be 3.8 million t. With adequate supplies, no maize imports were permitted in 1999 beyond the 53,250 t that Thailand had committed to allow under the World Trade Organization agreement. In contrast, soybean output was about 375,000 t in 1999, with consumption expected to increase marginally to 1.17 million t. This means that soybean imports will rise to 800,000 t. In addition, about 680,000 t of soybean meal was produced in 1999 – 100,000 t from local soybeans and the rest imported.

Over 50% of world soybean production is transgenic varieties, mainly from North America, despite regulations governing genetically modified organisms or GIOs becoming more and more restrictive. In mid-1999, for example, the European Agriculture Commissioners made a political agreement with regard to the ban on the use of GIOs in feed. As a net food producer, Thailand should be able to deal with potential problems. DNA analysis has been used to confirm the origin of raw materials used in food processing to comply with trade agreements. The DNA Fingerprinting Unit will check the species identification of tuna already canned. This addresses the conflict between global free trade and environmental protection. The US Department of Commerce proposes to inhibit the importation of Atlantic-caught bluefin tuna harvested from countries using methods that are inconsistent with the International Convention for the Conservation of Atlantic Tunas.

## References

Attathom, S. and Sriwatanapongse, S. (1994) Present status on field testing of transgenic plants in Thailand. In: *Proceedings of the 3rd International Symposium on the*

*Biosafety Results of Field Tests of Genetically Modified Plants and Microorganisms, Monterey, California*. pp. 349–352.

Attathom, S., Siriwong, P., Kositratana, W. and Sutabutra, T. (1990) Improvement of transformation efficiency of *Agrobacterium* mediated gene transfer in tomato. *Kasetsart Journal (Natural Science)* 25, 15–20.

Attathom, S., Sriwatanapongse, S. and Wongsatithorn, D. (1996) *Biosafety Capacity Building: Evaluation Criteria Development*. Stockholm Environment Institute, Sweden, pp. 59–60.

Chaopongpang, S., Mahon, R., Poonpipit R., Srathonghoy, K., Kositratana, W., Bateson, M., Burna, T., Attathom, S. and Dale, J. (1996) Transformation of Thai papaya (*Carica papaya* L.) with the coat protein gene of papaya ringspot virus via particle bombardment. Paper presented at the Third Asia Pacific Conference on Agricultural Biotechnology: Issues and Choices, Hua Hin, Prachuapkirikhan.

Disthaporn, S. (1994) Current rice blast epidemics and their management in Thailand. In: Zeigler, R.S., Leong, S.A. and Teng, P.S. (eds) *Rice Blast Disease*. CAB International, Wallingford, UK, pp. 333–432.

Flegel, T.W. (1997) Special topic review: major viral diseases of the black tiger prawn (*Penaeus monodon*) in Thailand. *World Journal of Microbiology and Biotechnology* 13, 433–442.

Phaosang, T., Ieamkhaeng, S., Bhunchoth, A., Patarapoowadol, S., Chiemsombat, P. and Attathom, S. (1996) Direct shoot organogenesis and plant regeneration from cotyledons of pepper (*Capsicum* spp.). Paper presented at the Third Asia Pacific Conference on Agricultural Biotechnology: Issues and Choices, Hua Hin, Prachuapkirikhan.

Sriwatanapongse, S., Tanticharoen, M. and Bhumiratana, S. (2000) Thailand. In: Tzotzos, G.T. and Skryabin, K.G. (eds) *Biotechnology in the Developing World and Countries in Economic Transition*. CAB International, Wallingford, UK, pp. 168–175.

Withyachumnarnkul, B., Boonsaeng, V., Flegel, T.W., Panyim, S. and Wongteerasupaya, C. (1998) Domestication and selective breeding of *Penaeus monodon* in Thailand. In: Flegel, T.W. (ed.) *Advances in Shrimp Biotechnology*. National Centre for Genetic Engineering and Biotechnology, Bangkok.

Yuthavong, Y. (1987) The impact of biotechnology and genetic engineering on development in Thailand. *Journal of the Science Society of Thailand* 13, 1–13.

Yuthavong, Y. (1999) An overview of biotechnology and biosciences in Thailand. *Thai Journal of Biotechnology* 1 (1), 1–11.

# Africa

# 8

# Egypt

Magdy A. Madkour

| | |
|---|---|
| Area (km²) | 1.001m |
| Cropland | 2% |
| Irrigated cropland | 32,460 km² |
| Permanent pasture | 0% |
| Population (1999 est.) | 67.2m |
| Population per km² | 67 |
| Annual population growth rate (1999 est.) | 1.82% |
| Life expectancy (men) | 60 yrs |
| (women) | 64 yrs |
| Adult literacy | 51.4% |

| | |
|---|---|
| GDP (1998 est.) | US$188 bn |
| GDP per head (1998 est.) | US$2850 |
| Annual growth in real GDP (1998 est.) | 5% |
| Agriculture as % of GDP | 20% |

Major export commodities:
crude oil, cotton, textiles

Major commodities:
cotton, rice, maize, wheat, bean, fruit, vegetables, meat

## Summary

Biotechnology research is promoted in Egypt through national research institutes. The Academy of Scientific Research and Technology (ASRT) is the principal institute responsible for biotechnology research programme development across all sectors. The Ministry of Agriculture and Land Reclamation has specific responsibility for agricultural applications.

The Agricultural Genetic Engineering Research Institute (AGERI) was established in 1990 for advanced research in agricultural biotechnology. It is a component of the Ministry of Agriculture's Agricultural Research Centre (ARC), which has responsibility for agricultural research. AGERI has developed training programmes and focuses on the development of biotechnology for potato, rape-seed and tomato.

In 1990, ASRT proposed setting up a National Institute for Genetic Engineering and Biotechnology, to focus on basic and applied industrial biotechnology. The national institute was intended to complement applied expertise with advanced research on biochemistry and genetic engineering.

## Introduction

Agriculture plays a significant role in the national economy of Egypt, so it receives high priority from government. Agriculture accounts for 20% of gross domestic product (GDP) and total exports and 34% of the total labour force. The agricultural sector contributes to the overall food needs of the country and provides the domestic industry with agricultural raw materials.

The agricultural sector has taken major steps to reform its economic policy programme, such as the following:

- Gradually removing government controls on farm output prices.
- Increasing farm-gate prices to cope with international prices.
- Removing farm input subsidies.
- Removing government constraints on the private sector in importing and exporting agricultural crops.
- Imposing limitations on state ownership of land and sale of new land to the private sector.
- Adjusting the land tenancy system.
- Confining the role of the Ministry of Agriculture (MOA) to agricultural research, extension and economic policies.

As the government moves towards privatization, transfer of technology to the private sector is also occurring (for example, *in vitro* micropropagation of virus-free potato). This shows the capacity and interest of the private sector in adopting new technology. Technology transfer is expected to grow dramatically in the short term as the research programmes become more product-orientated.

One of the major targets for biotechnology in Egypt is the production of transgenic plants conferring resistance to biotic stresses (resulting from pathogenic viruses, bacteria, fungi and insect pests) and abiotic stresses, such as salinity, drought and high temperature. These biotic and abiotic constraints are major agricultural problems, leading to serious yield losses in many economically important crops in Egypt.

The Agricultural Genetic Engineering Research Institute (AGERI) was established in 1990 at the Agricultural Research Centre (ARC) to promote the

**Table 8.1.** Examples of current plant genetic engineering research at AGERI/Egypt.

| Discipline | Potato | Tomato | Cotton | Maize | Faba bean | Cucurbits | Wheat | Banana | Date-palm |
|---|---|---|---|---|---|---|---|---|---|
| Virus resistance | ✓ | ✓ | | | ✓ | ✓ | | ✓ | |
| Insect resistance | ✓ | ✓ | ✓ | ✓ | | | | | |
| Stress tolerance | | ✓ | ✓ | | ✓ | | ✓ | | |
| Genome mapping and fingerprinting | | ✓ | | ✓ | | | | | ✓ |
| Fungal resistance | | ✓ | | ✓ | ✓ | | | | |

transfer and application of this new technology. AGERI aims to adopt the most recent technologies available worldwide to address problems facing agricultural development (Table 8.1) (see also Komen, 2000; Moawad and Madkour, 2000).

The agricultural sector has the following strategic goals:

- Optimizing crop returns per unit of land and water consumed.
- Enhancing sustainability of resource use patterns and protection of the environment.
- Bridging the food gap and achieving self-reliance.
- Expanding foreign-exchange earnings from agricultural exports.

Some of the opportunities for deploying modern biotechnological approaches include:

- Producing transgenic plants resistant to indigenous biotic and abiotic stress.
- Reducing the use of agrochemicals and pesticides and their environmental risks.
- Improving the nutritional quality of food crops.
- Reducing the dependence on imported agricultural products (seeds, crops).

The private sector has access to biotechnology and has invested heavily in research and development (R & D) of technology and the necessary expertise to bring a product to market. The competitive edge of a private company depends on the proprietary nature of its R & D and the protection offered by intellectual property laws. A private company might engage in development of a product in conjunction with a developing country because: (i) it addresses a technical problem critical to its own product development, (ii) it presents an opportunity to enhance its public relations, and/or (iii) it provides a window on an important market, technology or germ-plasm of interest. Developing-country institutions may be interested in working with private companies to gain access to important technology, develop managerial and business expertise, build intellectual capacity or form a partnership with an entity that has an existing capability of bringing a product to market.

## Opportunities

### Pioneer Hi-Bred/AGERI: a private/public partnership

The relationship between AGERI, an Egyptian public-sector institution and Pioneer Hi-Bred, a US private company, was forged through common business interests. The importance of codevelopment of technology as opposed to technology transfer is especially pertinent in the case of Pioneer Hi-Bred's relationship with AGERI. In this partnership, a public-sector institution was able to bring a significant contribution to the table. AGERI has isolated a number of strains of *Bacillus thuringiensis* (Bt) that have pesticidal activity of interest to a private-sector company. AGERI filed a patent in Egypt on 11 January 1996 (No. 019797) on a Bt-derived bioinsecticide against a wide range of insects and in the USA on 10 January 1997 (No. 5178–3) on Bt isolates with broad-spectrum activity.

AGERI also has a state-of-the-art biocontainment facility and a team of trained scientists. AGERI can provide access to the local Egyptian market and the broader Middle East market, both of which are sufficiently developed to be attractive. In turn, Pioneer Hi-Bred came to the discussion table with technology and with marketing, regulatory and legal expertise of value to AGERI.

AGERI therefore initiated a partnership through a US Agency for International Development (USAID) R & D grant to achieve the following:

- Research and training for AGERI scientists to be trained at Pioneer/Iowa in methodologies relating to agricultural biotechnology.
- Potential for product development: Pioneer was granted access to evaluate certain novel Bt proteins and genes patented by AGERI.

Joint discoveries resulting from the work will be shared and patent rights will be sought according to the terms of the USAID agreement. Pioneer Hi-Bred will retain sole ownership of its proprietary Bt gene(s) and proprietary germplasm and AGERI will retain sole ownership of its proprietary Bt gene(s).

Under a separate agreement, AGERI granted material transfer agreements (MTAs) for Pioneer Hi-Bred to evaluate the Bt toxin protein. Options for possible commercial development of maize have also been considered. This is one of the examples of USAID-sponsored collaboration as described by Lewis (2000).

### BIOGRO/AGERI: a business partnership

A second model of moving research into commercial application is demonstrated through the successful interaction between scientists at AGERI and the University of Wyoming, who have been involved in collaborative research studies for the past 6 years on Bt. The research efforts led to the development of a biological pesticide based on a highly potent strain of Bt isolated from the Nile Delta. This strain is extremely effective against a broad range of insects:

Lepidoptera (moths), Coleoptera (beetles) and Diptera (mosquitoes). An additional significant feature of this strain is its capacity to kill nematodes.

AGERI has successfully managed to manufacture its first biopesticide, Agerin, based on Bt. Agerin is capable of protecting a broad range of important agricultural commodities and of controlling a number of biomedically significant pests and has the potential for sales on a worldwide scale.

To fulfil its commitments to bring research results into application and large-scale commercial distribution to farmers, AGERI, in collaboration with a private investor, succeeded in establishing a commercial business entity under the name BIOGRO International. This company is responsible for the commercialization of research results conducted in AGERI and will be in a position to sell AGERI products. This is essential in order to guarantee that sales revenue generated from product sales will be invested back into the institute to support the continuation of its activities.

It is envisaged that BIOGRO will link with the Genetic Engineering Services Unit (GESU), which was established at AGERI to work out any commercial agreements to benefit both the institute and BIOGRO. It will also allow for the free flow of information and products related to genetic engineering being produced by the institute for commercialization.

AGERI attaches a high priority to collaborating with the private sector, which will be fully informed of R & D in the field of genetic engineering and biotechnology in Egypt through: (i) circulating newsletters and reports; (ii) having representatives of the private sector participate in the design of product R & D; and (iii) including a representative of the private sector on the governing board of AGERI.

As one of the leading institutions in agricultural genetic engineering in West Asia, North Africa and the Middle East, AGERI is planning to share its know-how and experience with other countries within the framework of Technical Cooperation among Developing Countries (TCDC). This will be achieved through specialized workshops, seminars and internships. The institute can also provide professional consultation in molecular biology and agricultural genetic engineering.

## International Cooperation

The international agricultural research centres (IARCs) could usefully expand their activities in the following areas to further assist national institutes in the applications of modern biotechnology

### Biosafety

- Setting up regional linkages to share biosafety data and to pool information.
- Providing training and guidance on risk assessment and risk management

issues.
- Providing technical training in biosafety reviews, prior to releases.
- Building consensus among nations on biosafety protocols and guidelines.
- Assisting in the development of media and information materials to increase public awareness.

### R & D collaboration

- Increasing Consultative Group on International Agricultural Research (CGIAR)/national agricultural research systems (NARS) collaboration in biotechnology R & D.
- Setting up programmes for the use and management of technology.

### Intellectual property management

- Increasing awareness on intellectual property and its fundamentals (copyrights, trade marks, patents, licensing, plant variety protection and plant breeders' rights).
- Establishing intellectual property policy and institutional policies.
- Building capacity and human resources development in the field of technology transfer and intellectual property rights.

## References

Komen, J. (2000) International initiatives in agri-food biotechnology. In: Tzotzos, G.T. and Skryabin, K.G. (eds) *Biotechnology in the Developing World and Countries in Economic Transition.* CAB International, Wallingford, UK, pp. 15–32.

Lewis, J. (2000) Leveraging partnerships between the public and private sector – experience of USAID's Agricultural Biotechnology Program. In: Persley, G.J. and Lantin, M.M. (eds) *Agricultural Biotechnology and the Poor: Proceedings of an International Conference, Washington, DC, 21–22 October 1999.* Consultative Group on International Agricultural Research, Washington, DC, pp. 196–199.

Moawad, H. and Madkour, M. (2000) Egypt. In: Tzotzos, G.T. and Skryabin, K.G. (eds) *Biotechnology in the Developing World and Countries in Economic Transition.* CAB International, Wallingford, UK, pp. 77–82.

# 9

# Kenya

N.K. Olembo

| | |
|---|---|
| Area ($km^2$) | 582,650 |
| Cropland | 7% |
| Irrigated cropland | 660 $km^2$ |
| Permanent pasture | 37% |
| Population | 28.8m |
| Population per $km^2$ | 43 |
| Ann. pop. growth rate (1999 est.) | 1.59% |
| Life expectancy (men) | 46.6 yrs |
| (women) | 47.5 yrs |
| Adult literacy | 78% |
| GDP (1998 est.) | US$43.9 bn |
| GDP per head | US$1500 |
| Growth in real GDP (1998 est.) | 1.6% |
| Agriculture as % of GDP | 29% |
| Value of agricultural exports (1998) | US$660m |
| Agricultural products as % of total exports | 33% |
| Major export commodities: | tea, coffee, maize, horticulture |
| Major subsistence commodities: | maize, millet, sorghum, wheat |

## Summary

The national institutes involved in agriculture-related research are the: Kenya Agricultural Research Institute (KARI), Kenya Trypanosomosis Research Institute (KETRI), Kenya Forestry Research Institute (KEFRI) and Kenya Marine Fisheries Research Institute (KEMFRI).

KARI reviewed agricultural activities under its mandate to establish priority areas for commodity research. Animal production (dairy and beef) occupies the highest priority for research. Among crops, maize, brassicas, dry bean and Irish potato are key commodities.

Biotechnology is being applied in the search for vaccines and diagnostics for some of the most important animal diseases.

Kenya's Industrial Property Office, established under the Industrial Property Act, deals with patent issues, including those relating to biotechnology.

## Introduction

Throughout the 1980s and 1990s, there has been considerable discussion on the importance, relevance and scope of biotechnology in developing countries. Likely areas of interest and priority have been considered and the suitability of the different biotechnologies in solving agricultural problems debated (Bureau of Science and Technology, 1988; Doyle, 1988a; Sasson, 1988; KARI/GTZ, 1989; Ketchum, 1989; Sasson and Costarini, 1989; KARI, 1990, 1991; Persley, 1990a, b; Brouwer *et al.*, 1991; Persley and Doyle, 1999; Persley and Lantin, 2000; Serageldin and Persley, 2000).

The problems to be solved and the areas of priority must be determined by those directly concerned with agricultural research and rural development in individual countries. Country studies, seminars and workshops held in individual countries are important to provide sufficient information for planners for identification of what is required for the successful applications of biotechnology.

The potential for biotechnology in Kenyan agriculture was investigated, including a review of national priorities and policies and the current status of biotechnology applications. Information was obtained through discussions with various government officials, policy-makers and scientists from national and international institutions in Kenya. Ndiritu (2000) gives a general overview of biotechnology in Africa, with emphasis on the Kenyan situation.

As used here, biotechnology includes traditional biotechnology based on conventional techniques, such as those used in rapid micropropagation by tissue culture, food fermentation processes and conventional animal vaccine production, as well as modern biotechnology, including the techniques of recombinant DNA technology, monoclonal antibody production and new cell and tissue-culture techniques.

### Science and technology policy

The importance of science in development has been recognized in Kenya for many years. This realization prompted the creation of the Ministry of

Research, Science and Technology in 1985 and the National Council for Science and Technology (NCST) in 1977.

The meaning, scope and applications of biotechnology are increasingly affecting Kenya's policy-makers. Much has still to be done to inform appropriate government officials about the nature and advantages of biotechnology applications. The Ministry officials see a need to promote biotechnology in the various institutions, at both the research and the production levels. The Ministry of Research, Science and Technology has frequently commented on the need to incorporate and promote biotechnological activities in Kenyan research. At the start of the 1990s, Kenya had a number of externally funded projects with a biotechnology component (Table 9.1).

The NCST is the coordinating body for science and technology, including policy issues on biotechnology. A focal point for coordination is the National Biotechnology Advisory Committee, under the Ministry of Research, Science and Technology, established by the Minister of Research, Science and Technology in 1990. It is composed of the directors of the research institutes under the Ministry (covering agriculture, industry, health and the environment). The committee has been expanded to include representatives from the private sector. Its purpose is to advise the Minister on policy and institutional issues in the use of biotechnology across all sectors in Kenya (Hassanali, 2000).

It is envisaged that the advisory committee will be responsible for such policy issues as patents, biosafety, ethics and commercialization of inventions and discoveries. The NCST has also established a Secretariat to look into intellectual property issues and to develop guidelines on biosafety.

## Institutional issues

The issue of priorities in agriculture and how biotechnology was to be promoted at the national level was discussed in two meetings in Kenya in the late 1980s and early 1990s (KARI, 1989; Mailu *et al.*, 1991).

The May 1989 workshop (KARI, 1989) was prompted by issues discussed at the Third Conference of the International Plant Biotechnology Network (IPBNet) held in Nairobi in January 1989. This seems to have been the beginning of biotechnology awareness, not only for Kenya, but also for the other interested African countries. It led to the establishment of the African Plant Biotechnology Network (APBNet) and highlighted to the participants the advantages of biotechnology techniques in research and production.

In both national meetings, identification of priority areas was stressed. The 1990 conference developed priorities in plant and animal biotechnologies (see later sections).

Another common recommendation from both meetings was the need to set up a Biotechnology Centre. Although the 1989 workshop participants

**Table 9.1**. Examples of externally funded projects with a biotechnology component in Kenya.

| Institute | Project | Funding |
|---|---|---|
| Kenya Agricultural Research Institute (KARI) | Small Ruminant Programme, DNA probes | USAID |
| Kabete Veterinary Lab., KARI, Muguga | Vaccine research | DFID, UK |
| KARI Plant Quarantine Station | Gene bank | Germany |
| KARI, Thika | Propagation of pyrethrum, potato tissue culture | FAO |
| | Horticultural crops development | UNDP |
| Dept. of Toxicology and Pharmacology, University of Nairobi | Salmonella, amoeba identification, DNA probes | Norwegian Agency for International Development (NARAD) |
| Dept. of Soil Science, University of Nairobi | Rhizobium inoculant | UNESCO, UNEP, FAO, International Atomic Energy Agency (IAEA), Commonwealth Fund for Technical Cooperation (CFTC), Commonwealth Science Council (CSC) |
| CIP | Potato tissue culture | Consultative Group on International Agricultural Research (CGIAR) |
| ILRAD (now ILRI) | Vaccine for theileriosis, identification of trypanosomes, DNA probes, monoclonal antibodies | CGIAR |
| ICIPE | Ticks, tsetse-fly control: monoclonal antibodies, biological insecticides | Finnish International Development Agency (FINNIDA), IAEA, USAID |
| Jomo Kenyatta University College of Agriculture and Technology | Tissue culture of horticultural crops | Japan International Cooperation Agency (JICA) |
| KEFRI | Tissue culture of trees | JICA |
| KETRI | Identification of trypanosomosis, DNA probes, monoclonal antibodies, antigen ELISA | IAEA |

USAID, US Agency for International Development; DFID, Department for International Development; FAO, Food and Agriculture Organization; UNDP, UN Development Programme; UNESCO, UN Educational, Scientific and Cultural Organization; UNEP, UN Environment Programme; CIP, Centro Internacional de la Papa; ILRI, International Livestock Research Institute; ICIPE, International Centre of Insect Physiology and Ecology; ELISA, enzyme-linked immunosorbent assay.

suggested KARI as the possible host for the centre, the 1990 meeting participants took the view that it should be a more autonomous establishment, with a broader management style, and that it should involve all national aspects of plant and animal biotechnologies. Debate on these issues continues.

## Agricultural Sector

Agricultural production in Kenya is carried out by both large-scale (> 19 ha) and small-scale farmers. Large-scale commercial farming, a practice carried over from the colonial period, thrives both under individual private land ownership and through government parastatals such as the Agricultural Development Corporation (ADC). More than 50% of the land held by large-scale farmers is utilized as meadows and pastures for livestock production (Table 9.2). A large portion of this land falls under the medium- or low-potential category. Only a small part (12%) of large-scale farming is under crop production and this is in the high-potential areas.

Approximately 4% of the total agricultural land is used for large-scale farming and this is mainly in the high- to medium-potential areas (Republic of Kenya Statistical Abstracts, 1990).

Small-scale farmers are of two types: (i) those cultivating traditionally inherited smallholdings in rural Kenya (2 million farms, 2 ha or less); and (ii) those in resettlement schemes on land acquired by the government from previous colonial landowners (80,000 farms, 2–7 ha) (Durr and Lorenzl, 1980). Most small-scale farmers own the land they cultivate. Land held by them totals approximately 40 million ha, covering both high- and low-potential areas. Mixed farming is practised to a great extent, and food and cash crops are grown. Pastoral tribes, essentially nomadic, practise communal land ownership and primarily occupy medium- to low-potential lands.

Kenya relies heavily on cash crops to generate foreign currency. Agricultural produce forms 33% of all domestic exports, most of which is accounted for by a few major crops: coffee 20%; tea 27%; pyrethrum 1.7%;

**Table 9.2.** Land utilization of large farms (1988) (from Republic of Kenya Statistical Abstracts, 1990).

| Land utilization | '000 ha | % of total |
|---|---|---|
| Temporary crops | 90.2 | 3.5 |
| Permanent crops | 198.0 | 8.0 |
| Temporary meadows | 45.0 | 1.2 |
| Uncultivated meadows | 1310.1 | 52.7 |
| Pastures | 95.2 | 3.8 |
| Forests | 98.5 | 4.0 |
| All other land (fallow) | 650.5 | 26.2 |

and horticultural crops and pulses 22% (Republic of Kenya Statistical Abstracts, 1990).

Food crops such as maize, millet, sorghum, cassava, potato and pulses form the main staple diets and must be produced in ever-increasing quantities to satisfy the demands of a rapidly growing population. The annual population growth rate of Kenya was estimated to be 1.59% in 1999. These facts, compounded by the diminishing land space available for farming, make it imperative that Kenya adopt more intensive methods of farming to increase its agricultural production.

## Agricultural Research

Kenya has six national research institutes under the Ministry of Research, Science and Technology. Four of these (KARI, KETRI, KEFRI and KEMFRI) have agricultural production mandates, including forestry and fisheries.

KARI carries the largest load of agricultural responsibilities and operates through 17 national research centres (NRCs), located mainly in the west-central part of the country.

There are two major constraints in agricultural production in Kenya: (i) agroecological constraints, caused by diseases, pests and weeds, as well as climatic and soil conditions; and (ii) institutional and socio-economic constraints, including infrastructure and access to inputs and markets.

### Research priorities

An intensive exercise was carried out in KARI in 1991 to establish priority areas for research for both crop and livestock production. The priority-setting was based on a scoring system, which considered factors such as value of production, export earnings, import saving, food security, employment potential and future expansion potential (KARI, 1991). Using this method, commodity and research priorities were established for 53 items, the first 20 of which are:

1. Dairy
2. Beef
3. Shoat
4. Maize
5. Brassica
6. Dry bean
7. Irish potato
8. Sugar cane
9. Banana
10. Poultry
11. Wheat
12. Pyrethrum
13. Tomato
14. Cotton
15. Floriculture
16. Onion
17. Asian vegetables
18. Mango
19. Sorghum
20. Groundnut

**Table 9.3.** Major crops and their important pests and diseases (from ICIPE, 1989; KARI, 1990).

| Crops | Diseases | Pests |
|---|---|---|
| Cereals<br>Maize, wheat, sorghum, rice, millet, barley | Maize streak virus, leaf blight, rusts, cob rot, smut | Stalk borer, army worm, aphid, cutworm, shoot fly, crickets, leafhoppers, termites |
| Industrial crops<br>Coffee, tea, sugar cane, pyrethrum, cotton | Coffee-berry disease, sugar-cane mosaic virus, nematode, cutworm, bacterial blight | Coffee borer, bollworm, coffee scale |
| Vegetables<br>Brassicas, tomato, onion, carrot | Nematodes, wilt, leaf spot, early blight, spider mite, white fly, *Xanthomonas campestris* | Spider mite, white grubs |
| Fruits and flowers<br>Banana, mango, citrus, avocado, pawpaw, passion, flowers | Leaf spot, nematodes, greening, powdery mildew | Red spider mite |
| Roots and tubers<br>Irish potato, cassava, sweet potato, yam | Potato blight, wilt, *Erwinia*, cassava mosaic | Mealy-bug, weevils, whitefly, greenfly |
| Grain legumes<br>Dry bean, French bean, pigeon-pea | Wilt, bean common mosaic virus, rust | Cutworm, spider mite |
| Oil-seeds<br>Groundnut, sunflower, coconut, *simsim*, soybean | Leaf spot, wilt, rust, leaf blotch | Spider mite, hoppers, bollworm, whitefly |

The major crops and their important pests and diseases are listed in Table 9.3. Maize, brassicas, dry bean, Irish potato, sugar cane, banana, wheat, pyrethrum, tomato and cotton are key crops, in that order of priority. Coffee and tea were not assessed in this exercise, since they occupy special status, being managed under the Coffee Research Foundation and the Tea Research Foundation, respectively. The major livestock species and their important diseases are given in Table 9.4.

*Food crops*

Maize is the most important staple in the Kenyan diet. It is widely grown, both at subsistence and commercial levels. Maize research programmes are in progress at the KARI national research stations at Kitale and Katumani. These deal with breeding and agronomy, utilizing conventional and new methods.

Current constraints to maize production are: drought tolerance (to extend

**Table 9.4.** Major livestock species and their important diseases in Kenya.

| Livestock species | Disease | Diagnosis | Vaccine available | Treatment |
|---|---|---|---|---|
| Cattle | Tick-borne diseases | Serological | Infection/ | Parvaquone |
| | East Coast fever | DNA probes | treatment | Halofuginone |
| | | Monoclonal antibodies | method | |
| | Babesiosis | Serological | – | Imazol |
| | Anaplasmosis | DNA, serological | Yes | Tetracyclines |
| | Cowdriosis | Serological | – | Not effective |
| | Viral diseases | | | |
| | Foot-and-mouth disease | Serological | Yes | Not effective |
| | Rinderpest | Serological | Yes | Not effective |
| Cattle, sheep and goat | Lumpy skin disease | Serological, clinical | Yes | Not effective |
| | Malignant catarrhal fever | Serological | No | Not effective |
| Cattle, sheep, goats, domestic animals | Rabies | Clinical, serological | Yes | Not effective |
| Sheep | Nairobi sheep disease, blue tongue | Serological | Yes | |
| Sheep, cattle, goats | Rift Valley fever | Serological | Yes | |
| | Helminthosis | Clinical | No | Antihelminthiosis |
| Cattle, sheep, goats, pigs | Cysticercosis | Clinical | No | |
| | Hydatidosis | Clinical | No | |
| | Nematodosis | Clinical | No | |
| | Fasciolosis | Clinical | No | |
| Sheep and goats | Haemonchosis | Clinical | | |
| Cattle, sheep, goats, pigs, camels | Trypanosomosis | Serological, DNA probes Monoclonal antibodies | No | Berenil Samorin |
| Cattle | Contagion-borne | Serological | | |
| | pleuropneumonia (CBPP) | Serological | Yes | Antibiotics |
| Goats | Contagious caprine pleuropneumonia (CCPP) | Serological | Yes | Antibiotics |
| Cattle, poultry | Brucellosis | Serological | Yes (imported) | |
| Cattle, canine | Leptospirosis | Serological | Yes (imported) | Antibiotics |
| Cattle, sheep, goats | Anthrax | Serological | Yes | Not effective |
| Poultry | Fowl typhoid | | Yes | |
| Poultry | Newcastle disease | | Yes | |

Source: KARI Animal Health Section.

the maize-growing regions beyond the highland areas with regular rainfall); maize streak virus; stalk borers; and storage pests.

Some of these constraints may be solved using conventional breeding techniques. For example, resistance to maize streak virus is available and could be incorporated into Kenya's élite maize hybrids through conventional breeding.

Other constraints may benefit from the application of modern biotechnology. Considerable effort is being put into the use of restriction fragment length polymorphism (RFLP) mapping technique in maize breeding in North America and Europe. The use of RFLPs would enable the specific targeting of desirable characteristics for Kenyan maize varieties.

Research on sorghum and millet is conducted at the KARI research centres at Kitale and Katumani. Production of these traditionally drought-tolerant crops has declined in recent years, and yet the need to utilize marginal lands has become increasingly important. The key constraints are drought and salinity and these are the main targets of existing breeding programmes. New varieties with better baking quality would also stimulate production.

Potato, sweet potato and cassava are popular staple foods. Sweet potato and cassava are major crops in low-rainfall areas. Research into these crops is conducted at the KARI research centres at Kakamega, Katumani and Mtwapa. Potato research is carried out at Tigoni (in collaboration with the International Potato Centre (CIP)). The major constraints to production are diseases, often leading to the lack of good-quality planting material. The major pests and diseases affecting cassava, potato and sweet potato are listed in Table 9.5.

New biotechnologies could be potentially useful in the production of dis-

**Table 9.5.** Major pests and diseases of root and tuber crops in Kenya (from Ewell and Kirkby, 1990; KARI, 1990).

| | Diseases | Pests |
|---|---|---|
| Cassava | Cassava mosaic virus | Cassava mealy-bug (*Phenacoccus manihoti*)<br>Cassava green mite (*Mononychellus tanajoa*) |
| Potato | Viral diseases<br>Late blight (*Phytophthora infestans*)<br>Stem blight (*Alternaria* spp.)<br>Bacterial wilt (*Ralstonia* (formerly *Pseudomonas*) *solanacearum*)<br>Root diseases | Potato tuber moth (*Phthorimaea operculella*) |
| Sweet potato | Viral diseases | Sweet potato weevil (*Cylas puncticoliae*, *Cylas brunneus*, *Cylas farmicarius*) |

ease-free planting material. Various methods of tissue culture should be undertaken for rapid multiplication and disease elimination in cassava, potato and sweet potato. Potato production is under high viral infection pressure and additional micropropagation for viral elimination is urgently needed. This would reduce dependence on imported healthy potato planting material, as is currently done, and would reduce importation costs. Similar technologies are required for sweet potato. Rapid micropropagation of cassava is required to provide sufficient planting material at appropriate times of the year. Demand for cassava material is over 1 million plants, although conventional methods can supply only 20% of requirements. Drought-tolerant cassava varieties are also required.

Although large quantities of horticultural crops are consumed locally, a lucrative export market exists for most of these products. There is, therefore, a great demand for planting material, which cannot be met with current conventional methods of propagation.

Floriculture is also an important export earner; it is at present dominated by a few large commercial growers/companies. There is limited access for small-scale farmers into the highly competitive export trade. Useful biotechnologies for horticultural crops include:

- Tissue culture for mass production of clean planting material of selected species (e.g. banana, strawberry).
- Selection of disease-resistant material, particularly virus infection.
- Selection for resistance to *Phytophthora* spp. for citrus and avocado.
- Tissue culture of ornamentals.

Tissue culture is an important factor in horticultural production in Kenya. The technique has been successfully applied at the research and production level to a number of horticultural crops, such as citrus, passion-fruit, strawberries, sugar cane, Irish and sweet potatoes, ornamentals and banana.

Commercially, biotechnology application in horticulture has potential. In many cases, efforts have been lacking to utilize technologies for increased horticultural production for the small-scale farmer. The multiplication of potato material in the CIP programme is distributed through the ADC. Large-scale micropropagation activity for banana is undertaken through an ISAMA-sponsored programme.

For such horticultural crops as strawberries, citrus and ornamentals, tissue-culture techniques are available at the Department of Crop Science of the University of Nairobi. However, the services of this department must be strengthened in order to supply farmers with the much-needed disease-free planting material.

The Plant Quarantine Station at Muguga has for several years been involved in tissue-culture work, directed towards micropropagation of potato (in conjunction with CIP), cassava, sweet potato, strawberries, hops and ornamentals. Extension of these techniques to the National Gene Bank is envisaged in the future.

Pulses, such as beans, pigeon-pea, cow-pea, groundnut and lentils, form an essential part of the Kenyan diet. Beans are widely grown and popularly consumed (18 kg per capita year$^{-1}$) (Ewell and Kirkby, 1990).

The major constraints limiting production of pulses include: poor soil fertility, diseases (e.g. bacterial blight, *Xanthomonas phaseoli*, bean common mosaic virus, anthracnose, *Colletotrichum lindemuthianum*) and insects (e.g. bean fly, *Ophionyia* spp.; weevils, bruchids, *Acanthoscelides obtectus*).

Potentially useful biotechnologies for pulse production include:

- Improved nitrogen fixation systems for low soil fertility. Rhizobia, legumes and soil mineral utilization are ways of combating low soil fertility. Rhizobia have been found to greatly increase legume production after inoculation of seed with the appropriate symbionts (MIRCEN, 1986). Mycorrhiza associations greatly increase the rate of uptake of nutrients, particularly phosphorus and nitrogen.
- Exploitation of these biological systems should be encouraged and current production levels of rhizobia inoculant increased, for effective distribution to farmers. More research is needed into mycorrhiza associations and identification.
- Pest and disease resistance.
- Biological control of insect pests.
- Drought and salinity tolerance, to broaden the scope of production.

The Nairobi *Rhizobium* programme is one of the UN Educational, Scientific and Cultural Organization (UNESCO)-sponsored Microbiological Resources Centre (MIRCEN) projects located in five continents. Africa hosts three MIRCENs, situated in Dakar, Cairo and Nairobi. The MIRCEN programme in Nairobi is managed by the Departments of Soil Sciences and Botany of the University of Nairobi.

Rhizobia culture collection, preservation, testing and inoculant preparation are the major activities of the programme. So far, 216 bacterial strains have been collected locally and from other centres for preservation. Routine preparation of inoculant for various grain pasture and tree legumes is carried out. In 1990, 1500 kg of inoculant was produced, most of which was for the common bean (40%), lucerne (23%), soybean (14%) and *Desmodium* (9%).

Results of trials conducted at Kabete and Embu in 1979 comparing five strains of *Rhizobium leguminosarum* biovar. *phaseoli* showed that inoculation was better than the application of 40 kg N ha$^{-1}$ and in the case of one strain, Nitragin, bean yields were as good as those given 80 kg N ha$^{-1}$ (Anaynaba, 1981). In other trials involving four strains of *Bradyrhizobium japonicum*, soybean yields as high as 3000 kg ha$^{-1}$ were obtained. In effect, most strains fixed more nitrogen than could be supplied by the addition of 90 kg N ha$^{-1}$ (Kena and Imbamba, 1986).

Current research priorities are focused on screening for tolerance to high temperatures, soil acidity (in relation to phosphorus deficiency), drought and salinity. Local strains of *R. leguminosarum* biovar. *phaseoli*, which can tolerate high temperatures (as high as 42°C), have been isolated. This is significant in the preparation of heat-resistant and stable inoculant suitable for use in very hot areas.

Over 20 strains of *Bradyrhizobium* species isolated from groundnut nodules are being tested for biological competence and tolerance to low pH and low phosphorus. This is important, since phosphorus deficiency is a limiting factor in nitrogen fixation.

Other areas of future research will focus on more effective methods of culture characterization through the use of molecular biology techniques. Research in the departments of biochemistry and botany of the University of Nairobi attempted to characterize *Rhizobium* DNA from the various strains and prepare probes for identification of the different cultures. Future application of recombinant DNA techniques could provide means of culture improvement through genetic manipulation.

In addition to the use of *Rhizobium* inoculants as biofertilizers, the MIRCEN team is exploring the possibilities and effectiveness of mycorrhiza in enhancing uptake of phosphorus and water by plants. The use of mycorrhiza inoculum would thus be highly beneficial in poorer soils, such as those in most of Kenya's agricultural areas.

Although production and supply of *Rhizobium* inoculant have steadily increased from 40 kg in 1981 to 1500 kg in 1990, they are far below the quantity required to make any significant impact on the total production of leguminous crops in Kenya. This is possibly due to lack of marketing skills in the programme. Although experimental results have shown the efficacy of inoculation, by which yields are invariably higher than those receiving commercial nitrogen fertilizers, most farmers are unaware of the advantages of inoculant application.

### *Cash crops*

*Coffee.* Research on coffee is conducted at the Ruiru Coffee Research Foundation. Coffee in Kenya is highly susceptible to two major diseases: coffee-berry disease (CBD) and coffee-leaf rust (CLR). Together these diseases can cause yield losses of over 70%.

The introduced hybrid Ruiru II combines resistance to CBD and CLR. It is high-yielding and of good quality and saves about 30% of input to produce a ton of coffee. To avoid segregation of these properties in propagation through seed production, vegetative propagation of coffee would be valuable. Present methods of micropropagation are too slow to meet the demand for planting material of the new clone. Clearly a faster method of coffee propagation is required.

Tissue culture has been initiated at the Ruiru Station. It will be important to establish a rapid clonal propagation system from single plants in order to avoid segregation.

At the Ruiru Coffee Research Foundation, a small team of scientists is involved in developing tissue-culture methods for micropropagation purposes.

It has been possible to induce somatic embryogenesis in secondary calluses of leaf plants. Meristem culture is also being attempted. Callus regeneration is a problem and methods sidestepping this stage, such as direct embryogenesis, should be tried.

*Tea.* Kenya has a well-established tea-breeding programme at Kericho. Conventional methods have been successfully applied in producing adequate planting material. New biotechnological methods will be required in identification of élite traits in existing material and their incorporation into breeding clones.

Rapid propagation of élite clones via meristem culture or somatic embryogenesis would take advantage of existing variability in yield and growth vigour to supply the farmer with better planting material.

*Sugar cane.* In the past 20 years, Kenya has given enormous attention to the production of sugar cane in its efforts towards self-sufficiency in sugar supplies. Sugar-cane mosaic virus is a major constraint, causing huge losses due to clonal decline. New biotechnologies for the rapid production of clean planting material would be useful.

*Pyrethrum.* Demand on the world market for natural pyrethroids extracted from pyrethrum has shown a remarkable increase in recent years. Kenya enjoys more than 80% of this market. Pyrethrum production, which at one time almost ceased under the threat of synthetic derivatives, has made a comeback. This has placed considerable strain on the supply of planting material. Micropropagation methods, based on explanted axillary buds, are being utilized to alleviate the situation. Other tissue-culture methods, such as somatic embryogenesis, should be explored to increase the propagation rate. Other constraints are diseases and nematodes.

Tissue culture has been applied successfully to the propagation of pyrethrum in Kenya. This involves the culture of axillary buds, a technique that yields 2500 plantlets in 2 months from a single axillary bud, as compared with fewer than 50 over the same period through the conventional splitting method (Okioga *et al.*, 1989). Considering that world demand in the early 1990s for pyrethrum (15,000 t of dried flowers) far outstripped Kenya's production (8000 t), the need for rapid propagation was apparent. Over 1.5 million plants had been produced in this way by 1992, but the demand from farmers has yet to be met. Consequently, the Molo National Pyrethrum Research Centre is expanding its capacity for tissue culture.

The key attraction of the micropropagated plants to pyrethrum farmers lies in the cleanness of the material, which is free from nematodes and disease. The resultant plants show vigorous growth, with higher flower yield, compared to conventionally propagated plants.

*National Gene Bank*
New biotechnological techniques are required at the National Gene Bank at Muguga. The gene bank, established in 1987, has experienced a growing demand for collection, multiplication, rejuvenation and storage of germ-plasm over the the past decade.

*Forestry/agroforestry*
Rapid depletion of Kenya's indigenous forests and the severity of the encroaching desert conditions over the arid and semi-arid areas have prompted the government to take measures to promote reafforestation in all parts of the country. The establishment of KEFRI and the more recent creation of the Ministry for Reclamation and Development of Arid, Semi-arid and Wastelands is an indication of the priority being given to these problems.

Among the programmes at KEFRI, the biotechnology research programme is quickly becoming essential in developing reafforestation systems for Kenya's arid and semi-arid lands. The programme is also geared towards the promotion of farm forestry and agroforestry.

To achieve these objectives, tissue-culture and legume *Rhizobium* technologies have been developed and are being applied. For example, *in vitro* micropropagation is being applied for tree species whose seeds are either unobtainable (e.g. *Ocotea usambarensis* – camphor wood), of low viability (e.g. *Grevillea robusta* – silky oak) or difficult to propagate by conventional methods (e.g. *Cordeuxia edulis* – Yeheb nut – and *Chlorophora excelsa* – mvule).

*Grevillea robusta* is one of the most useful and most popular of the exotic tree species in Kenya. Imported into the country from Australia early in the 20th century as a shade tree for coffee and tea, it has acquired many different roles, including utilization in agroforestry, as a fodder crop, wood fuel and timber. As a result, demand has outstripped supply, bringing into focus the need for *in vitro* propagation to produce enough plantlets for farmers. Trials on the KEFRI compound at Muguga have demonstrated the rapid growth of *in vitro*-propagated plants as compared with controls, probably as a result of disease elimination.

Similarly, *C. excelsa*, the most popular hard timber tree in East Africa, is receiving attention. Extensive exploitation has almost depleted this valuable tree, prompting its protection by Presidential Decree. Micropropagation using meristematic buds has produced rooted plantlets that have been transferred to the field. It is thus possible that the East African Mvule will once again flourish. Another hardwood of magnificent beauty as timber is the Elgon teak (*Dalbizia egotzi*), a mahogany species of western Kenya. Its slow growth has prevented production of adequate supplies for afforestation purposes. *In vitro* propagation has been achieved and plantlets have already been distributed to some farmers.

Other tree species being multiplied through micropropagation are *O. usambarensis*, *C. edulis* and *Eucalyptus granolis*. This latter species, when micropropagated, has grown three times faster than it does in Australia, its country of origin. It can be harvested as a wood fuel in less than 2 years and as timber in 4 years in Kenya.

Perhaps the greatest advantage of tissue culture in Kenya's forestry activities will be realized when it comes to the rescue of Kenya's pine forests, currently under threat from aphid attack. Identification of resistant species is being conducted and subsequent mass propagation of resistant clones should provide material for reafforestation. This approach has already been demonstrated for *Pinus radiata*, a good pulp and paper tree, which was decimated by *Dothistroma pini*. Identification of resistant species has led to micropropagation of disease-tolerant clones, which may once again form the much needed pulp-tree forests.

The *Rhizobium* project at KEFRI is geared towards maximum exploitation of the beneficial tree microorganisms for improvement of tree productivity and soil fertility. Seventy-five *Rhizobium* strains have been identified and characterized, including strains symbiotic to *Acacia albida*, *Acacia mearnsii*, *Calliandra calothyrsus*, *Leucaena leucocephala*, *Sesbania grandiflora* and *Sesbania sesban*, among others. Appropriate inoculants are now routinely prepared and sold cheaply to farmers. Laboratory tests have shown much faster growth of inoculated plantlets as compared with controls applied with commercial fertilizer. This has clearly been demonstrated with the *Acacia* and *Sesbania* tree species, whose growth rate has been shown to be more than twice the rate of controls.

Collaboration with the International Centre for Research in Agroforestry (ICRAF), both in the mass propagation of suitable tree species for agroforestry, such as *C. grevillia*, and in the utilization of *Rhizobium* inoculants for improved productivity, is desirable for maximum exploitation of these technologies in Kenya's forestry activities.

The significance of the utilization of cheap biofertilizers such as these must not be underrated for Kenya. The country's dependence on imported fertilizers can be considerably reduced through exploitation of alternative sources. Currently Kenya imports more than 18 million kg of fertilizer per year. Increased use of biological fertilizers would use considerable foreign exchange. The added advantage of nitrogen fixation and phosphorus mobilization by soil organisms is evident in their capacity to improve soils generally and thus contribute to overall food production.

## Plant and Animal Biotechnology Priorities

A national conference on plant and animal biotechnology was held in Nairobi from 25 February to 3 March 1990 (Mailu *et al.*, 1991). Information from this conference and more recent consultations with people working in national institutions and government ministries show that the question of priorities for biotechnology in Kenya is being investigated and at several levels. The key issue is not whether biotechnologies should be applied, but rather whether the areas for application have been properly identified. Priorities for agricultural biotechnologies for Kenya were enumerated both for plant and animal production. The priorities were to:

- Develop tissue-culture procedures for use in propagation and pathogen elimination in root and tuber crops, maize cultivars, horticultural crops, oil crops and other food and cash crops of economic importance.
- Utilize non-conventional methods in selecting desirable traits in cell cultures.
- Develop diagnostic methods for the detection of plant pathogens, particularly viral, bacterial and fungal pathogens. Techniques to be utilized would include monoclonal-antibody and recombinant DNA probes.
- Conserve vegetatively *in vitro* and distribution of germ-plasm of ordinary plants as well as recalcitrant species.
- Develop techniques such as fingerprinting and RFLP for use as molecular markers in plant breeding and selection.
- Transfer of useful genes into plants to develop pest and disease resistance.
- Increase of nitrogen fixation ability in symbiotic microbes such as rhizobia.
- Identify, propagate and utilize indigenous plants (trees, vegetables) of potential use to Kenya.
- Use microbes in bioconversion of natural materials, which may be applied as plant fertilizers.
- Develop methods of biocontrol for insect pests and diseases.

## Animal production and health

### *Constraints analysis*

Kenya places high priority on livestock development, because animal products are an important source of protein for most of the population. Milk, meat, fish and poultry are an essential part of traditional diets. Domestic animals continue to play major roles in the social activities of various tribes, including their importance as an income source. Thus nearly every rural household in Kenya has some livestock. The numbers (and species) vary, depending on availability of land, pasture, social practices and disease distribution.

Large-scale commercial livestock production began a century ago. The introduction of higher-yielding exotic breeds, particularly cattle, sheep and poultry, has contributed immensely to increased livestock production. Few small-scale farmers are able to keep exotic stock, because of the low survival rates caused by the vulnerability of introduced species to diseases.

Even when zero grazing of cattle is practised and regular dipping undertaken as a means of prevention of tick-borne diseases, high losses still occur. Small-scale farmers therefore prefer to keep the more resistant local cattle, in spite of their lower productivity. Disease eradication and effective vaccination programmes would clearly boost livestock production in Africa.

The important livestock diseases are listed in Table 9.4. The endemic diseases (Doyle, 1988a; KARI, 1990, 1991) pose the greatest challenge in terms of diagnosis, treatment and eradication (Table 9.6). Information obtained from the Veterinary Laboratories Services at Kabete shows the need to intensify research into better diagnostic methods and more effective vaccines, particularly for those diseases that cause high mortality rates and for which chemotherapy is either unattainable or too expensive for smallholders.

*Vaccines and diagnostics*

Monoclonal-antibody and recombinant DNA techniques offer vast scope in the diagnosis of many livestock diseases (Doyle, 1988a; Persley, 1990a, b; Swedish Council for Forestry and Agricultural Research, 1990). Cheap, accurate and easy-to-use diagnostic methods would not only be directly advantageous to the farmer, but would also provide quick and accurate tools for epidemiological data required for national prevention and eradication programmes.

Vaccine development for some of the diseases listed in Table 9.4 presents special problems. The search for effective vaccines against trypanosomosis and theileriosis is a long-term research priority, which is being investigated by the International Livestock Research Institute (ILRI), among others. Promising results are being obtained by ILRI for the development of a vaccine to combat theileriosis and trypanosomosis.

**Table 9.6.** Endemic diseases of livestock in Kenya, their treatment, diagnosis and prevention (from KARI Animal Health Section; KARI, 1990).

| Disease | Vaccine availability | Current diagnostics | Treatment |
|---|---|---|---|
| Rinderpest | ✓ | Serological | None |
| Trypanosomosis | ✓ | Serological<br>MCA, DNA* | Berenil,<br>Samorin |
| Theileriosis | ✓ | Serological | Clexon |
| Foot-and-mouth disease | ✓ | Serological | Not effective |
| Redwater disease | ✓ | Serological | Not effective |
| Blackwater fever | ✓ | Serological | Not effective |
| Lumpy disease | ✓ | Serological | Not effective |
| Swine fever | ✓ | Serological | Not effective |
| Rift Valley fever | ✓ | Serological | Not effective |
| Anaplasmosis | ✓ | DNA, serological | Not effective |
| Nairobi sheep virus disease | ✓ | DNA | Not effective |
| Newcastle disease | ✓ | DNA | Not effective |
| Anthrax | ✓ | DNA | Not effective |

*Methods developed at the International Laboratory for Research on Animal Diseases, ILRAD (now International Livestock Research Institute (ILRI)), KETRI.

### *Livestock breeding*

Livestock breeding in Kenya is conducted at stations in Naivasha and Kikuyu and by the National Insemination Programmes. These programmes continue to utilize conventional methods in the improvement of local breeds and the introduction of exotic types. Many other countries are using embryo transfer and embryo splitting and developing transgenic animals (Bureau of Science and Technology, 1988; Sasson and Costarini, 1989; Persley, 1990b; Swedish Council for Forestry and Agricultural Research, 1990). These technologies have yet to be widely used in Kenya. Although the slowness of technology transfer in animal biotechnology may be attributed to high costs, a more likely explanation is that the introduction of exotic breeds still has to overcome the barrier of disease and nutritional inadequacies. It is argued that there is no point in striving to introduce a super-breed when existing stock cannot be properly fed and disease still poses such huge threats.

Constraints for Kenya must therefore be tackled first and the improvement of nutrition and prevention of disease are prerequisites to the successful development of exotic breeds.

### *National animal biotechnology conference priorities*

In the KARI/US Agency for International Development (USAID) Biotechnology conference in February 1990 (Mailu *et al.*, 1991), participants listed the following as priority areas for biotechnological application in animal production and improvement in Kenya:

- Development of diagnostic tools using monoclonal-antibody and recombinant DNA probes for identification of disease in animals.
- Development of vaccines through biotechnological means for the eradication of diseases of economic importance to animal production in Kenya.
- Identification and cloning of genes for disease resistance and also genes that may confer superior reproductive performance in animals.
- Development of embryo-culture systems, embryo cloning and *in vitro* fertilization.

### *Current status of livestock biotechnology applications*

Vaccine development for livestock in Kenya focuses mainly on the important diseases of cattle, sheep, goats and poultry.

At the national level, impressive work is in progress at the Kabete Veterinary Laboratories, where several projects have been initiated in the biotechnology section, focusing on small ruminants. The project is part of the Small Ruminant Programme launched in developing countries some years ago and supported by USAID.

In one project, work on *Babesia bigemina* (redwater disease), which affects cattle, has determined the mode of infection of the red blood cells of the host animal. Protein molecules deposited on the surface of the host cell by the invading organism have been isolated, purified and characterized. Monoclonal antibodies have been prepared against some of the proteins and purification of the protein antigens through immunoaffinity techniques is under way. This will produce enough antigen for further investigations into the possibilities of a vaccine.

Although current statistics are not available, a 1972–1974 Food and Agriculture Organization (FAO) report gives the rate of infection of cattle as 50.4%. Considering that redwater disease is almost always fatal, this research should receive closer attention and support.

Another interesting project at Kabete involves the development of virus-vectored recombinant vaccines. Here, the research targets the use of capripox virus to deliver genes coding for protective viral antigens into host animals. Current work involves the identification of antigens and their role in protection against various viral infections, such as Rift Valley fever, babesiosis and the Nairobi sheep virus disease. Once the antigen genes have been identified, insertion into capripox virus would provide a multipurpose virus-vectored vaccine. A suitable insertion system is currently being investigated.

A project already completed is one involving the development of a DNA probe for *Anaplasma ovis* in goats (Shompole *et al.*, 1989). The high sensitivity and specificity of the probe (sensitivity of up to 35% parasitaemia) makes the probe ideal for use in epidemiological studies. It has, in fact, been utilized to determine the disease prevalence in various parts of Kenya, where infection ranges between 22% and 80%. Production of a kit would be the next step, but this would require commercialization.

Vaccine development efforts for agricultural animals in Kenya have focused mainly on diseases of cattle, sheep, goats and poultry. The most prevalent diseases in cattle are theileriosis, trypanosomosis, foot-and-mouth disease, blackwater fever and lumpy disease. Theileriosis and trypanosomosis together cause annual losses of 3 million cattle.

Work at ILRI endeavours to develop a vaccine against *Theileria parva*, the causative agent of theileriosis. Scientists at ILRI have identified, characterized and isolated genes coding for two sporozoite antigens. One of the antigens found on the surface of sporozoites from different parasite strains induces antibody formation in cattle (Doyle, 1988b) and may be utilized for the preparation of a cross-strain vaccine. Several other antigens are under similar investigation in order to develop a robust vaccine.

ILRI has well-developed molecular technologies for identification of antigens and monitoring of the immune response. Monoclonal-antibody and recombinant DNA technologies are also well established at ILRI. The International Centre of Insect Physiology and Ecology (ICIPE) is tackling the menace of East Coast fever from a different approach. Tick gut components are used to prepare protein antigens that produce host resistance to the parasite through induction (ICIPE, 1989). Methodologies involved include immunological techniques for the identification, isolation and characterization of the antigen. However, large-scale production of antigens has yet to be embarked upon, a necessary prelude to vaccine development.

Eradication of these tropical diseases should be a target for the Kenyan government, if the more productive exotic animal breeds are to be introduced successfully into rural Kenya. Indeed, the government's aims are well demonstrated by the establishment of a parastatal body, the Kenya Veterinary Vaccine Production Institute, charged with the responsibility for vaccine development, production and marketing.

Availability of cheap, accurate and easy-to-use diagnostic methods is one of the requirements for livestock disease treatment and eradication in Kenya. Without these, accurate epidemiological information will not be available to assist in outbreak surveillance. Indeed, this has been a constraint in livestock disease management in rural Kenya.

At ILRI and KETRI, diagnostic methods based on recombinant DNA and monoclonal-antibody techniques have been developed for the diagnosis of trypanosomes, the causative agents of sleeping sickness in people and nagana (trypanosomosis) in animals.

At ILRI, probes have been developed that can distinguish between *Trypanosoma congolense* and *Trypanosoma simiae*, morphologically identical parasites but which cause different diseases in livestock (Doyle, 1988b). Further development of the technique will offer accurate means of identification of pathogens that are closely related morphologically.

ICIPE has recently initiated work on *Bacillus thuringiensis* as part of its biological control programme for insect pests and vectors (ICIPE, 1989). Pathogens isolated include those showing specific toxicity to: *Chilo partellus* (stem borer), *Spodoptera exempta* (African army worm), *Busseola fusca* (stalk borer), *Glossina* (tsetse-fly) and *Aedes* (mosquito).

Toxin isolation, purification and characterization from the different *B. thuringiensis* strains is in progress. The significance of *B. thuringiensis* bacteria as biopesticides comes in the wake of worldwide concern for environmental degradation caused by excessive utilization of commercial pesticides.

## Training

Training in biotechnology is required for existing staff and to create a pool of people with these sophisticated skills. Training may be short term, in the form of workshops or more intensive courses, or it may take several years, resulting in higher degrees.

National agricultural research institutions, including KARI and KETRI, have in the last few years sent employees for advanced training. It is expected that several of these people will be exposed to biotechnological techniques applicable to their areas of research.

KARI's training programme is supported by various international donors. In recent years there has been concerted effort by sponsors to support training at the postgraduate level. For short-term programmes, KARI has received sponsorship from the International Service for National Agricultural Research (ISNAR) and the European Economic Union (EU) among others.

Promising programmes suffer from inadequate facilities. Once strengthened, their contribution to personnel development in biotechnology should be substantial. National universities in Kenya have also initiated programmes. The key issue in supporting local training is that skills are transferred to larger numbers with comparatively less expenditure per capita than is achieved with overseas training.

The multidisciplinary nature of biotechnology requires parallel development of other disciplines such as biochemistry, immunology, molecular biology, virology and microbiology. Training in these fields is well catered for by the various departments of national universities in Kenya, supplemented by overseas opportunities. Retention of staff in Kenyan institutions will be a critical requirement for the success of training programmes. This requires motivation in the form of adequate remuneration, suitable physical facilities and a good work environment. The government should introduce policies that ensure proper management of human resources in the field of biotechnology, to avoid the loss of qualified staff as happens in many African countries.

The international agricultural research centres (IARCs) could play a greater role in national training programmes by providing facilities to nationals within their region of operation. ILRI and ICIPE have offered training facilities for limited numbers of scholars at the MS and PhD levels within the region. The excellent facilities at ILRI match any international requirements. Limitations exist, however, where mandates have to be observed and flexibility becomes limited. ILRI can handle only a very few trainees at a time. Several Kenyan scientists have received postgraduate training at ILRI.

The African Regional Postgraduate Programme in Insect Science (ARPPIS) programme at ICIPE deserves mention. ICIPE took the lead many years ago in providing facilities for expert training for Africa in its areas of specialization. Several Kenyan scientists have received training through the programme. Its role in biotechnology training will become more apparent as ICIPE establishes its units.

Another useful undertaking would be for international institutes in Africa to conduct short-term workshops geared towards specific techniques in biotechnology.

The creation in June 2000 of the Doyle Foundation (in honour of the late Dr J.J. Doyle, Deputy Director General of ILRAD (now ILRI), who spent 20 years in animal research in East Africa) will offer further training and other opportunities for African scholars and others (www.doylefoundation.org).

## Information and Documentation

There is a considerable amount of information on agricultural biotechnology, but it is not reaching all target groups, including policy-makers, researchers, producers and farmers. Information dissemination in developing countries is often haphazard and slow and may not be in a form easily understood by the target group.

Kenya, like most African countries, has poor information systems. Effective access to international computer databases has yet to be established. A computerized network linking local institutions with such databases would be a desirable undertaking.

Although library services have expanded in recent years, the number of journal subscriptions has decreased, due to high cost and foreign-exchange limitations. Even local universities have had to cut down on subscriptions, causing a serious drop in information sources. For biotechnology, it means that new journals have no chance of appearing in local libraries.

International institutions such as ILRI, ICIPE and ICRAF may play a significant role in allowing use of their library services, but national systems have a duty to ensure adequate information diffusion to target groups.

Several African networks have been formed, essentially for information dissemination and related issues:

- African Biotechnology Stakeholders Forum (ABSF).
- African Plant Biotechnology Network (APBNet).
- African Biosciences Network (ABN).
- African Association of Insect Scientists (AAIS).
- African Network of Scientific and Technological Institutions (ANSTI).
- African Centre for Technology Studies (ACTS).
- Association of Faculties of Science of African Universities (AFSAU).
- Association for the Advancement of Agricultural Sciences in Africa (AAASA).
- African Academy of Sciences (AAS).
- ISAAA Afsi Center

Most of these networks disseminate information through journals, newsletters, seminars and conferences.

Some programmes have inbuilt information services. For example, the MIRCEN programme in Kenya publishes a biennial newsletter and also reaches farmers through demonstrations in national agricultural shows.

National institutions and associations are making noticeable advances through various symposia and seminars to communicate scientific information within Kenya.

## Patent Issues

There are two regional organizations in Africa that deal with patent issues (Kameri-Mbote, 1991):

**1.** Organisation Africaine de la Propriété Intellectuelle (OAPI), to which the francophone countries belong.
**2.** African Regional Intellectual Property Organization (ARIPO), based in Harare and to which most of the anglophone countries subscribe.

ARIPO members are divided into three categories:

**1.** Botswana, Lesotho and Swaziland, which confer automatic protection to patents registered in South Africa.
**2.** The Gambia, Ghana, Seychelles, Sierra Leone and Uganda, which require that patents be first registered in the UK before registration with the home country.
**3.** Kenya, Liberia, Malawi, Mauritius, Nigeria, Somalia, Sudan, Tanzania, Zambia and Zimbabwe, which have established independent patent laws. These either are based on guidelines provided by the World Intellectual Property Organization (WIPO) or utilize the patent laws of the UK as a model.

Kenya's Industrial Property Act, which was established in 1989, is partly patterned along the lines of the WIPO provisions. Clauses reflecting national requirements, such as environmental conservation, are incorporated.

The most important provision for biotechnology is that discoveries or innovations that are biotechnological are patentable under the law. Such inventions are excluded from the WIPO model. The emphasis on environmental conservation lends recognition to the need for utilization of biotechnology for sustainable development.

The Industrial Property Act removes obstacles contained in technology transfer procedures that restrict developing countries from marketing imported technology. It also removes restrictions imposed on the use of transferred technology for marketing of commodities regionally. These clauses are upheld in WIPO patent laws and impinge on the freedom of countries to make lasting gains from technology transfer.

Kenya's Industrial Property Act has some limitations. The Act makes no provision for the safeguard of indigenous raw materials and does not give guidelines concerning the exportation of germ-plasm from Kenya. The question of ownership of indigenous germ-plasm is likely to feature prominently as developing countries become increasingly aware of the importance of biotechnology (Juma, 1987, 1991; Clark and Juma, 1991).

Kenya's patent law is a step in the right direction, as it forms a basis for the assurance of intellectual property ownership locally. This will stimulate innovations and activate production at many levels. Its effectiveness will depend largely on how seriously the goals are established.

To implement the Patents Act, the Kenya Industrial Property Office has been established and an operational secretariat is in place. Applications for patents have commenced and examination of agreements is in progress.

## Biosafety

Kenya now has a system for monitoring the research, testing and release of genetically engineered organisms. The existing regulatory laws for agricultural management, such as the Crop Production and Livestock Act (1926), the Plant Protection Act (1937), the Seed and Plant Varieties Act (1972), the Fertilizers and Animal Foodstuffs Act (1963) and the Animal Diseases Act, do not have clauses specific to recombinant DNA technology. However, they may be able to be suitably modified to take into account the new technologies.

The plant quarantine legislation includes legal authority to control and regulate movement of plants and plant material entering the country. This forms a starting-point in the establishment of regulatory measures for biotechnology. Its import requirements and certification of plant genetic materials for export or import are applied in the monitoring of living organisms and agents that may possibly cause damage to agricultural, forestry or fibre crops in the country.

No specific regulations govern the production and utilization of genetically engineered animal products. Vaccines and diagnostic methods derived through new technologies will probably adhere to current provisions made through the Food, Drugs and Chemical Substances and the Pharmacy and Poisons Acts. These acts do not have specific clauses for the regulation of genetically engineered products.

Other regulatory laws that are in operation are the Public Health Act, the Radiation Protection Act and the Pest Control Products Act. These laws have enforcement clauses to ensure that safety measures are adhered to in the handling and administration of dangerous drugs and chemicals. Licensing and inspection exercises are carried out by the relevant Boards established for this purpose.

The regulatory system in use in the USA and other Organization for Economic Cooperation and Development (OECD) countries to ensure safety and efficacy in handling biological organisms and products is essential for both public and environmental safety (Bureau of Science and Technology, 1988; Persley, 1990b; Swedish Council for Forestry and Agricultural Research, 1990). These experiences have been drawn upon in the development of the Kenyan regulatory systems (Doyle and Persley, 1996).

Modern biotechnology can benefit developing countries only if safety measures are undertaken both in individual institutions, where safe handling of material is observed, and at the national level, where guidelines on testing and release must be established. Regulations should not be so unreasonably stringent that advantages of biotechnology are obscured. For example, transgenic occurrence, particularly in plants, has been demonstrated as a natural occurrence and should not be looked upon as a new concept (Swedish Council for Forestry and Agricultural Research, 1990). Therefore, the fears and risk in handling genetically improved plants need not be overly exaggerated. Still, this does not remove the need for caution and careful management of biotechnology applications.

For Kenya, a suitable regulatory system should ensure now that the potential use of modern biotechnology in research and production is receiving serious attention. Procedures for laboratory practice and field trials must be in place to avoid unintended consequences.

## Industry Activities

In Kenya, participation of industry in R & D is minimal. This stems from the fact that multinational companies operating in Kenya have well-established R & D programmes in their home countries. Moreover, it is argued that minimal financial returns, resulting from low market demands for commodities, prohibit investment in R & D. Little financial support or collaboration in joint R & D ventures in Kenya has been forthcoming.

This situation is in contrast to what is experienced in industrial countries, where large sums of money in the private sector are apportioned for research and close collaboration occurs between industry and the public sector (Bureau of Science and Technology, 1988; Sasson and Costarini, 1989; Swedish Council for Forestry and Agricultural Research, 1990). For example, many research units at universities in some industrial countries are funded in this way.

For biotechnology utilization to succeed in Kenya, new products must be developed to the commercial production stage. There is a need, therefore, to stimulate the collaboration and involvement of industry at each stage of the R & D process.

A few companies have opened discussion with national institutions for collaboration in specific areas. For example, Kenya Breweries Ltd has concluded an agreement with KARI on the study of certain crops, such as hop. Kenya Seed Company intends to set up a tissue-culture laboratory to strengthen its R & D programme. A government institution, Kenya Veterinary Vaccine Production Institute, is to liaise closely with research institutes, such as KARI, so as to utilize research findings in product formulation and commercialization.

A number of parastatals are involved in agricultural activities, including production and marketing. The Coffee Board of Kenya supports research at the Coffee Research Foundation and is also the key body in the marketing of produce. Similar functions are undertaken by the Tea Board, the Pyrethrum Board and the Cotton Board of Kenya. Other parastatal organizations involved in the agricultural infrastructure include the National Irrigation Board, Tana

River Development Authority, Keno Valley Development Authority and Lake Basin Development Authority. Large-scale farming is carried out by the ADC whereas the Agricultural Finance Corporation supports agricultural activities through direct loans to farmers.

The closer involvement of the private sector in the development of biotechnology policies and processes would be an advantage in Kenya.

## Conclusions

Application of biotechnology to increase and improve plant and animal productivity is an absolute requirement for Kenya if it is to feed its rapidly increasing population and utilize its limited land resources adequately.

There is need to evaluate national priorities and to determine which problem areas can best be solved by use of biotechnological techniques. The priority-setting exercises undertaken by KARI, for example, is a step in the right direction.

The areas regarded as most likely to benefit from biotechnology immediately are:

- Crop improvement through tissue culture for elimination of disease and for the supply of adequate, clean, planting materials to farmers (e.g. banana).
- Enhancement of nitrogen fixation by use of symbiotic microbes (biofertilizers).
- Development of diagnostic material by application of monoclonal-antibody and recombinant DNA techniques.
- Development of vaccines.

For the successful establishment of biotechnology in institutions, it is necessary that:

- Training be undertaken to produce high-calibre scientists.
- Incentives be offered to staff, particularly attractive salaries, to encourage retention of staff.
- Adequate facilities and infrastructure be provided for high productivity.
- Funding be available on a sustainable basis for the promotion of biotechnology utilization. In this regard, there is need to create awareness at the political and policy-making level of the advantages of biotechnology and thus the need for financial support for institutions.
- Biosafety standards and regulations need to be established.

Although there may be advantages in a national centre for biotechnology (Mailu *et al.*, 1991), its implementation does not seem feasible for the following reasons:

- Its establishment would require a government directive through the

Ministry of Research, Science and Technology, and would largely depend on the availability of funding for infrastructure and sustainable inputs.

- The current rate of training is too slow to produce the critical mass of capable scientists to staff the centre on a full-time basis. Scientists with biotechnological know-how currently available in various institutions are too few in number and are already heavily committed.

It is reasonable to suggest that decentralized biotechnology activities be promoted in relevant institutions. This will lead to more rapid adoption of biotechnological techniques, particularly if areas of priority are clearly identified. Such institutions may find it necessary to establish a central biotechnology laboratory for ease of funding and coordination of projects. However, the advantages of a national centre, which would ultimately provide greater coordination and training, should be given ongoing consideration.

## References

Anaynaba, A. (1981) Rhizobium and its utilization in African soils. In: Emajuaime, S.O., Ogumbi, O. and Sanni, S.O. (eds) *Global Impacts of Applied Microbiology, Sixth International Conference.* Academic Press, London, pp. 85–95.

Brouwer, H., Langeveld, H., Lursen, M. and Mugabe, J. (eds) (1991) *Biotechnology and Small-scale Agriculture in Kenya, an Inventory Study.* Department of Biology and Society, Vrije University, Amsterdam, the Netherlands, ACTS Press, Nairobi, Kenya.

Bureau of Science and Technology (1988) *Proceedings of the Meeting on Strengthening Collaboration in Biotechnology: International Agricultural Research and the Private Sector, Rosslyn, Virginia. 17–21 April, 1988.* Bureau of Science and Technology, Office of Agriculture, USAID, Washington, DC.

Clark, N. and Juma, C. (eds) (1991) *Policy Options for Developing Countries.* ACTS Press, Nairobi, Kenya.

Doyle, J.J. (1988a) *Current Status and Future Opportunities for the Use of Modern Biotechnology in Relation to Animal Health in Developing Countries.* Biotechnology Study Project Papers 3, World Bank, International Service for National Agricultural Research (ISNAR), Australian International Development Assistance Bureau (AIDAB) and Australian Centre for International Agricultural Research (ACIAR).

Doyle, J.J. (1988b) Current status and future opportunities for the use of modern biotechnology in relation to animal health in developing countries. In: *Proceedings of the Meeting on Strengthening Collaboration in Biotechnology: International Agricultural Research and the Private Sector.* Bureau for Science and Technology, Office of Agriculture, USAID, Washington, DC.

Doyle, J.J. and Persley, G.J. (eds) (1996) *Enabling Safe Use of Biotechnology: Principles and Practices.* ESD Monograph Series No. 10, World Bank, Washington, DC.

Durr, G. and Lorenzl, G. (eds) (1980) *Potato Production and Utilisation in Kenya.* Centro Internacional de la Papa, Lima, Peru.

Ewell, P.T. and Kirkby, R.A. (1990) Root tubers and beans in the food system in eastern and southern Africa. A paper presented at the Conference on Dialogue and

Training for the Promotion of Roots, Tubers and Legumes in Africa, 26–30 November.

Hassanali, A. (2000) Kenya. In: Tzotzos, G.T. and Skryabin, K.G. (eds) *Biotechnology in the Developing World and Countries in Economic Transition.* CAB International, Wallingford, UK, pp. 91–96.

ICIPE (1989) *Seventeenth Annual Report.* ICIPE Science Press, Nairobi, Kenya.

Juma, C. (ed.) (1987) *The Gene Hunters.* Princeton University Press, Princeton, New Jersey.

Juma, C. (1991) Biotechnology research in eastern Africa. Unpublished report.

Kameri-Mbote, P. (1991) Intellectual property and sustainable industrial development. A paper presented at the Workshop on Sustainable Industrial Development in Africa: Policy Perspectives for the 1990s. Nairobi, Kenya, 18–19 March, 1991.

KARI (1989) *Plant Biotechnology Workshop: Present and Future Biotechnology Research and Application for Kenya.* Workshop Summary. Nairobi, Kenya, 25–26 May, 1989.

KARI (1990) *Annual Report.* Kenya Agricultural Research Institute, Nairobi, Kenya.

KARI (1991) *Kenya's Agricultural Research Priorities to the Year 2000.* Kenya Agricultural Research Institute, Nairobi, Kenya.

KARI/GTZ (1989) *Plant Biotechnology. Studies on Present Status and Proposal for Future Application on Plant Tissue Culture and other Related Biotechnologies in Crop Improvement in Kenya.* Feasibility study between Deutsche Gesselschaft für Technische Zusammenarbeit (GTZ) and KARI, November. Kenya Agricultural Research Institute, Nairobi, Kenya.

Kena, S.O. and Imbamba, S.K. (1986) The East African Rhizobium MIRCEN. A framework for promoting regionally coordinated biological nitrogen fixation. *MIRCEN Journal* 2, 237–251.

Ketchum, J.L.F. (ed.) (1989) The Role of Tissue Culture and Novel Genetic Technologies in Crop Improvement. *Proceedings of the Third Conference of the International Plant Biotechnology Network (IPBNet), Nairobi, Kenya, TCCP, USAID, Colorado State University, 8–12 January, 1989.*

Mailu, A.M., Mugah, J.O. and Fungoh, P.O. (eds) (1991) Biotechnology in Kenya. *Proceedings of the National Conference on Plant and Animal Biotechnology, KARI, USAID, Nairobi, Kenya, 25 February–3 March, 1990.*

MIRCEN (1986) The Nairobi Rhizobium. *MIRCEN Journal* 2.

Ndiritu, C.G. (2000) Biotechnology in Africa: why the controversy? In: Persley, G.J. and Lantin, M.M. (eds) *Agricultural Biotechnology and the Poor: Proceedings of an International Conference, Washington, DC, 21–22 October 1999.* Consultative Group on International Agricultural Research, Washington, DC, pp. 109–114.

Okioga, D.M., Muriithi, L.M., Gichuru, S.P. and Ottaro, W.G.M. (1989) Rapid propagation of pyrethrum (*Chrysanthemum cinerariifolium vis.*) by tissue culture techniques. In: *Proceedings of the Third Conference of the International Plant Biotechnology Network (IPBNet), Nairobi, Kenya, 8–12 January.*

Persley, G.J. (1990a) *Beyond Mendel's Garden: Biotechnology in the Service of World Agriculture.* CAB International, Wallingford, UK, 155 pp.

Persley, G.J. (ed.) (1990b) *Agricultural Biotechnology: Opportunities for International Development.* CAB International, Wallingford, UK, 495 pp.

Persley, G.J. and Doyle, J.J. (1999) Overview brief. In: *Biotechnology for Developing Country Agriculture: Problems and Opportunities.* Brief 1 of 10. 2020 Vision Focus 2, International Food Policy Research Institute, Washington, DC 2 pp.

Persley, G.J. and Lantin, M.M. (eds) (2000) *Agricultural Biotechnology and the Poor:*

*Proceedings of an International Conference, Washington, DC, 21–22 October 1999.* Consultative Group on International Agricultural Research, Washington, DC 235 pp.

Republic of Kenya Statistical Abstracts (1990) Ministry of Planning and National Development, Central Bureau of Statistics, Nairobi, Kenya.

Sasson, A. (ed.) (1988) *Biotechnologies and Development.* UNESCO, Technical Centre for Agricultural and Rural Cooperation.

Sasson, A. and Costarini, V. (eds) (1989) *Plant Biotechnologies for Developing Countries. Proceedings of an International Symposium, CTA, FAO, Luxemburg, 26–30 June.*

Serageldin, I. and Persley, G.J. (2000) *Promethean Science: Agricultural Biotechnology, the Environment and the Poor.* Consultative Group on International Agricultural Research, Washington, DC, 48 pp.

Shompole, S., Waghela, S.D., Rurangirwa, F.R. and McGuire, T.C. (1989) Cloned DNA probes identify *Anaplasma ovis* in goats and reveal a high prevalence of infection. *Journal of Clinical Microbiology* 27, 2730–2735.

Swedish Council for Forestry and Agricultural Research (1990) *Proceedings of an International Conference on Advances in Biotechnology. Stockholm, Sweden, 1–14 March, 1990.* Swedish Council for Forestry and Agricultural Research, and Swedish Recombinant DNA Advisory Committee, Stockholm.

# Zimbabwe

**10**

C.J. Chetsanga

| | |
|---|---|
| Area ($km^2$) | 390,580 |
| Cropland | 7% |
| Irrigated cropland | 1930 $km^2$ |
| Permanent pasture | 13% |
| Population (1999 est.) | 11.1m |
| Pop. per $km^2$ | 28 |
| Ann. pop. growth rate (1997) | 1.26% |
| Life expectancy (1999 est.) (men) | 38.7 yrs |
| (women) | 38.9 yrs |
| Adult literacy | 85% |
| GDP (1998 est.) | US$26.2 bn |
| GDP per head | US$2400 |

| | |
|---|---|
| Growth in real GDP (1998 est.) | 1.5% |
| Ann. inflation (1998) | 32% |
| Agriculture as % of GDP | 18.3% |
| Value of agricultural exports | US$4.9 bn |
| Agricultural products as % of total exports | 30.7% |

Major export commodities: tobacco, cotton, sugar

Major subsistence commodities: maize, wheat, coffee, sugar cane, groundnuts, cattle, sheep, goats, pigs

## Summary

Some of the steps being taken to enable Zimbabwe to benefit from the new developments in biotechnology are reviewed. An important institutional development was the decision of the Government of Zimbabwe, at the recommendation of the Research Council of Zimbabwe (RCZ), to allocate funds for the establishment of the Biotechnology Research Institute (BRI). The BRI conducts research on agricultural, industrial and medical biotechnology. Another important development has been the launching of an MS programme in biotechnology at the University of Zimbabwe.

Guidelines for the regulation of biotechnology relating to the release of transgenic organisms into the environment have been prepared.

## Introduction

Agricultural production in Zimbabwe is concerned with both food crops for local consumption and cash crops for export. National policy is self-sufficiency in food production and maximizing export earnings by the efficient production of cash crops for export. Food and cash crops are grown under high inputs of fertilizer and irrigation because many regions of the country have poor soil and low and unpredictable rainfall. These inputs are costly (especially fertilizer) and not easily available to the small-scale farmer in village areas. There is a great need to develop agricultural approaches that can benefit smallholders in crop and animal production. Crops and livestock are often subjected to stress from pests and drought conditions. These factors impinge on the farmers' ability to increase agricultural output. These constraints have the potential of being alleviated by the wider application of biotechnology in Zimbabwe (Persley, 1990a, b; Chetsanga, 2000; Gopo, 2000; Persley and Lantin, 2000).

## Agricultural Policy

The government of Zimbabwe has a national policy that places food security as a top priority. As a result the government has generously funded agricultural research.

Modern biotechnology is a new area that is not yet well understood by many policy-makers. The level of national awareness of the potential contribution of agricultural biotechnology to food security in Zimbabwe needs to be rapidly enhanced.

The main vehicle for creating public awareness of issues in Zimbabwe is the press. Journalists need to be encouraged to write about agricultural biotechnology and its benefits to all Zimbabweans for the local newspapers and other media. The limited public understanding of the development of biotechnology helps explain why a national policy has not been pursued. Debate on safety and ethical issues relating to the applications of biotechnology is taking place in Zimbabwe.

The Research Council of Zimbabwe (RCZ) is developing guidelines and policy options for biotechnology (Woodend, 1995). It is expected that the establishment of a national policy on biotechnology will be accompanied by the allocation of more resources for agricultural biotechnology.

### Biotechnology programme

The development of biotechnology in Zimbabwe was spearheaded by RCZ, a government agency charged with promoting and directing the development of R & D and advising government about areas of new developments in science and technology (Anon., 1993).

The potential of biotechnologies in new industrial projects should make this an attractive area for public as well as private investment. Government should attach great importance to biotechnology because it has the potential for promoting the development of new industries.

The Zimbabwe biotechnology programme has been defined in broad terms by RCZ. In the National S&T Policy Statement, one of the major priority areas is agricultural biotechnology research for the improvement of crop and livestock production. The major thrust of the biotechnology programme as envisaged by RCZ includes efforts to develop high-yielding crops, food technology for nutritional needs, development of improved horticultural crops, improved methods of animal breeding and the creation of a gene bank.

The Zimbabwe Biotechnology Programme, especially its capacity-building component, has benefited from funding by the Dutch government, SAREC (Sweden) and the Rockefeller Foundation (Chetsanga, 2000).

## Biotechnology Research Institute

The implementation of the national programme will in part take place at the Biotechnology Research Institute (BRI), established under the Scientific and Industrial Research and Development Centre (SIRDC).

Personnel training has started at the University of Zimbabwe, where a number of students have graduated with an MS in biotechnology. Some of the recent graduates have been sent overseas for more specialized training, including agricultural biotechnology. They are now playing an important part in establishing a sustained research programme in biotechnology in Zimbabwe. There remains, however, an urgent need for many more trained agricultural biotechnology scientists. The private-sector research institutions also embrace biotechnology in their research activities.

## Regulatory Issues

Zimbabwean scientists and policy-makers have been working actively in developing guidelines for the safe use of biotechnology within Zimbabwe. Zimbabwe's Biosafety Regulations were gazetted as Statutory Instrument 1999 by the government. A Biosafety Board has been established to oversee the conduct of biotechnology in Zimbabwe (Chetsanga, 2000).

The rationale for developing national biosafety guidelines in Zimbabwe is to provide an 'enabling climate' that will facilitate the greater use of modern biotechnology by scientists in Zimbabwe (see Doyle and Persley, 1996). The existence of clear regulatory requirements will facilitate the development of technologies within the country and the import and evaluation of new biotechnological products developed elsewhere. Without such a regulatory framework, overseas collaborators, public and private, will be reluctant to make available new products (e.g. transgenic plants) for evaluation of their potential usefulness in Zimbabwe. The national regulatory framework for biotechnology should also be part of the raising of public awareness on biotechnology in Zimbabwe and reassure public interest groups that new technologies are being evaluated with due regard for any potential risks.

The private-sector groups that are establishing significant activities in genetic engineering have been concerned that there are no national guidelines and regulations dealing with recombinant DNA technology experimentation.

The underlying concern about the impact of genetically improved organisms (GIOs) is the perceived harmful effect that they might have on ecosystems. Despite thousands of field tests of GIOs, there has been no evidence to date to justify this concern. There is a need, however, to take all reasonable precautions to protect the environment and to ensure that no disruption occurs to the existing ecological equilibrium. It is also important that a regulatory system be flexible and be able to respond to accumulating experience in the use of GIOs worldwide.

## Patent Protection

Zimbabwe is a member of the African Regional Intellectual Property Organization (ARIPO), an intergovernmental organization. In 1982 the Council of ARIPO approved a Protocol on Patents and Industrial Designs, called the Harare Protocol. The Protocol empowers ARIPO to grant patents after a substantive examination and to register industrial designs on behalf of contracting states. By filing a single application, an applicant is able to designate several contracting states, thereby securing a monopoly protection within those states.

ARIPO conducts patent searches free of charge for contracting states, research institutions and individuals. Other services provided by ARIPO to member states include supplying patent information, staff training and offering advice on new technologies. It is essential to ensure that the intellectual property emanating from biotechnology research in the ARIPO contracting states will be licensed and protected under ARIPO.

The strong desire to commercialize new biotechnology by research organizations in industrial countries has made patenting such data routine procedure. Such organizations and their research staff are reluctant to invest in research in countries that do not have patent protection.

Developing countries need to weigh the need to offer intellectual property protection against not doing so, at the risk of losing opportunities for technology transfer. There remains the question of whether intellectual property rights are a 'right' that outweighs the need for food security in developing countries and whether these issues are mutually exclusive (see Lele *et al.*, 2000).

In Zimbabwe there is a recognition of the importance of offering intellectual property protection to the producers and sellers of new technologies and processes for both locally developed inventions and imported technologies. Zimbabwe will need to keep abreast of new developments so that it can harmonize its intellectual property protection for biotechnology with the international system.

## Government Incentives for Investment in Biotechnology

Although the government of Zimbabwe has not established a specific incentive mechanism for attracting investment in biotechnology, it is expected that the newly adopted national policy of trade liberalization will benefit biotechnology-based industries. The benefit would in turn stimulate the growth of the agroindustry sector.

A previous policy of restricting the proportion of profit that an overseas investor can repatriate has had the effect of discouraging international financiers from investing in Zimbabwe. The trade liberalization policy now allows overseas investors to repatriate a more generous proportion of the profits made on their investments in industries in the country.

It is expected that the ripple effect of trade liberalization will greatly benefit the development of biotechnology. The newly graduated Zimbabwean biotechnologists will also make it possible to establish viable biotechnology ventures in Zimbabwe. Investors always look at the availability of trained people in a particular country as an important prerequisite in establishing industrial activities in that country.

As biotechnology becomes more firmly established in Zimbabwe, it is possible that local biotechnology industries will spring up in specific sectors. The emergence of such industries will benefit from a newly formed Venture Capital Company of Zimbabwe (VCCZ), a resource that has been set up by government. In 1992 VCCZ assessed projects worth over Z$100 million (US$12.5 million). Access to this capital is strictly regulated. A group of assessors must first rule on the viability of a proposed project before it can be funded.

This means that the project proposers must do thorough research on the market and technological requirements of a venture. Such research is, of course, essential for biotechnology ventures, because the predictive parameters for success are still not well documented.

## Private-sector Activities

The private sector is a strong force on the Zimbabwe agricultural scene. This sector includes large-scale, commercial farmers and small-scale farmers. The large commercial farm group is well organized, with formalized structures for research in crop breeding and some horticulture. Some of the major commercial farmers' groups in Zimbabwe are:

- Commercial Farmers' Union
- Tea Growers' Association
- National Association of Dairy Farmers
- National Farmers' Association of Zimbabwe
- Commercial Grain Producers' Association
- Zimbabwe Tobacco Association
- Horticulture Promotion Council
- Commercial Cotton Growers' Association
- Oilseed Producers' Association
- Zimbabwe Sugar Association

Biotechnology activities of the private sector currently suffer from shortages of trained personnel. The sector has done well with traditional biotechnological approaches. It must now come to grips with the tools of the emerging modern biotechnology. The selection and evaluation of new biotechnologies by large-scale, commercial farmers and small-scale farmers will be essential for their wide-scale application.

Commercial farmers have been interested mainly in growing maize, cotton, soybean, fruit, tea, coffee and sugar cane. Some growers specializing in some of these crops have set up research laboratories to investigate their requirements for optimal growth. For example, the Zimbabwe Sugar Association has established a research station to study the problems involved in growing sugar cane under irrigation. Issues studied at the research station are varied, but include evaluation of many sugar-cane varieties, irrigation conditions, fertilizer requirements, disease susceptibility and pest control. A high priority is to identify drought-tolerant varieties of sugar cane. New biotechnologies, using marker-assisted plant breeding, may be able to assist in reaching this target.

Another example of the potential usefulness of biotechnology is to increase the efficiency of ethanol production from sugar cane.

## Agricultural Research

Organizations carrying out agricultural research in Zimbabwe include the: University of Zimbabwe; Department of Research and Specialist Services (DRSS); Tobacco Research Board; Cotton Research Board; Cattle Breeders' Association; and Pig Research Board. Among international organizations, the

International Crops Research Institute for the Semi-Arid Tropics (ICRISAT) and the International Centre for the Improvement of Wheat and Maize (CIMMYT) have programmes based in Zimbabwe.

The BRI focuses its services on commodities of interest to the small-scale farmers. BRI is conducting research on maize as its main priority, with sweet potato, mushroom and cassava being studied to exploit their potential as food and cash crops (Chetsanga, 2000).

The transnational companies operating in the agricultural sector in Zimbabwe are mainly those involved in producing fertilizers, pesticides and herbicides. These organizations rely on the research being done by their parent company.

## University activities

The University of Zimbabwe is the only one of the institutions of higher learning in Zimbabwe currently involved in agricultural teaching and research, including biotechnology.

The Faculty of Agriculture at the university was established in 1979. Its Department of Crop Science has established a teaching and research programme in modern agricultural biotechnology. A major project on cassava biotechnology is being implemented by the Department of Crop Science. A research programme on nitrogen-fixing bacteria (see below) has been launched in the Department of Soil Science and Agricultural Engineering.

These projects are complemented by medically orientated projects in the Departments of Biochemistry and Biological Sciences in the Faculty of Science. Projects in the Department of Biochemistry are focusing on hepatitis B virus, genome analysis and enzyme/protein technology.

## Government/industry cooperation

The area of greatest cooperation between government and industry in Zimbabwe is the support for agricultural research by government. The Ministry of Lands, Agriculture and Rural Resettlement has established the DRSS with laboratories to carry out agricultural research. At its central laboratories in Harare and research stations scattered throughout Zimbabwe, the DRSS undertakes research in crop and animal production, including veterinary services.

## Livestock production

Cattle, chickens and pigs are the most important animals raised in Zimbabwe. Sheep and goats are raised on a small scale. In recent years, the raising of game for meat is being promoted. The eland is a productive animal to raise for venison.

Animal science research includes beef and dairy cattle. A number of new animal breeds suited to local conditions have been developed. The Veterinary Services Department of DRSS undertakes research in livestock diseases and general animal health. This industry now enjoys the additional research services provided by the staff in the Faculty of Veterinary Sciences. This research includes work on animal virus and pest control.

Biotechnology can be used to improve domestic and wild animals. There is scope for developing animal breeds with resistance to a variety of pests. Artificial insemination can be an important technique in achieving this goal.

Research is needed to develop diagnostic probes for animal diseases. Livestock suffer from frequent attacks of foot-and-mouth disease. There is an urgent need to develop a national capability for early diagnosis and clinical management of the disease. It is expected that an increasing use of modern biotechnological protocols will be applied to many aspects of this research.

## Crop production

Tobacco growing has assumed a commanding lead in the area of crop improvement, followed by maize and cotton. Other grains grown are irrigated wheat, sorghum, *mhunga* and *rapoko*.

There is a growing oil-seeds industry, supported by the increasing national capacity to grow soybean, sunflower and groundnut. The industry is made up of a number of capable farmers, who will be watching the international technology scene with a view to adopting any new breakthroughs in crop-breeding methodologies.

The Crop Breeding Institute (CBI) at DRSS is involved in R & D to develop crop varieties with particularly attractive features for successful growth. A promising project is the crossing of wheat and rye to produce triticale. The hope is to use triticale as a substitute for wheat, most of which has been imported. Winter wheat is grown under irrigation.

Work on developing better maize varieties continues at CBI. There is a fruitful collaboration between CBI and CIMMYT in developing hybrid maize.

## Legume inoculant factory

DRSS has established a legume inoculant factory at the Grasslands Research Station in Marondera (70 km from Harare). The factory is providing rhizobium inoculant for commercial and communal farmers. Farming is often carried out in areas with poor soils and marginal rainfall. The use of commercial fertilizers in such regions can be costly, and wasteful if no rain falls after the fertilizer has been applied.

Rhizobium inoculant is cheaper and can be beneficially applied to groundnut, pea, soybean, bean, lucerne and sunhemp. The DRSS factory is steadily developing expertise and has the capability of developing inoculant technology and spreading it to other parts of Zimbabwe.

### University/industry cooperation

The area of cooperation between industry and university involves human resource training by the university. The Faculties of Agriculture, Science and Veterinary Science have played a central role in training people for agricultural biotechnology. This has included training in animal and plant breeding, soil fertility and pest and disease control, as well as the broad area of veterinary medicine. It also includes a practical training period in farm management, including fieldwork on several farms.

Horticulture has assumed greater importance in Zimbabwe agriculture in recent years, particularly through the transfer of conventional crop-breeding technologies. An increasing number of commercial farmers are growing horticultural plants. Some emerging horticulturists have received technical assistance from university experts and some small-scale farmers are gradually learning horticultural techniques.

An area in which there is an increasing level of cooperation between the university and industry is food science and technology. Because of the large size of the agricultural enterprise in Zimbabwe, there is enormous scope for the growth of a food-processing industry.

## International Cooperation

Zimbabwe scientists and policy-makers have benefited from attending workshops sponsored and organized by international organizations (e.g. Rockefeller Foundation, United Nations Development Programme, World Bank, International Service for National Agricultural Research). Some workshops have involved hands-on training in basic techniques in biotechnology and others have consisted of lectures and policy discussions.

In addition, the Australian, Dutch, Swedish and US governments have promoted the development of agriculture worldwide through some of their agencies. There are other governments and agencies that promote biotechnology without having a well-formulated aid policy for this specific area. Other smaller programmes include the International Science Programme in Chemical Sciences administered by Uppsala University in Sweden. This programme has offered fellowships to scientists from many Third World countries to acquire research training in chemical sciences (including biotechnology) in Sweden.

In Zimbabwe we have developed our own biotechnology training programme. We have implemented an MS biotechnology degree programme. This is jointly conducted by the Faculties of Agriculture and Science. The participants are trained in recombinant DNA technology, plant biotechnology, use of diagnostic-probe immunology, and downstream processing and protein technology. The course prepares students for careers in agricultural, industrial and medical biotechnology.

For this 2-year MS (Biotechnology), we were fortunate to obtain financial support from SAREC (Sweden) and the Netherlands Directorate General for International Cooperation (DGIS). We are confident that, after we have run several cycles of the course, we shall have a reasonable number of biotechnologists in the country. The MS course is also available to participants from other African countries.

BRI will also have research training programmes in agricultural, industrial and medical biotechnology. BRI will be able to provide short-term training for scientists from other countries in Africa.

## Conclusions

Biotechnology is one of the most promising instruments for promoting sustainable agriculture for socio-economic development. Its potential is steadily being more appreciated in Zimbabwean agricultural circles. However, the full realization of the potential benefit of biotechnology is being hampered by a number of constraints.

At the policy level, there is a lack of understanding of what biotechnology is, its benefits and its promotional requirements. Those who are keen to promote the greater use of biotechnology have not been as effective as they might be in raising the level of biotechnological awareness among the informed public and the decision-makers. There are still human resource constraints. We need Zimbabwean expertise in some of the areas of industrial biotechnology in which we wish to develop a focus in the MS programme. These areas include fermentation and downstream processing.

BRI will develop a comprehensive programme of activities, with laboratories for: plant biotechnology; animal biotechnology; food processing; horticulture; and medical biotechnology (including vaccine technology). Each one of these areas is quite extensive in its own right and will require specialized equipment, including a specially trained research staff.

It is hoped that assistance will be forthcoming from development agencies to purchase equipment and for training scientists for BRI. The availability of foreign currency to purchase research equipment and for training continues to be a major problem for the country.

## References

Anon. (1993) *Biotechnology Forum of Zimbabwe. Proceedings of the Workshop Prioritizing the Biotechnology Agenda for Zimbabwe. Harare, 11–12 May 1993.* BFZ/Enda Secretariat and Biological Sciences Department, University of Zimbabwe, Harare, Zimbabwe.

Chetsanga, C.J. (2000) Exploitation of Biotechnology in Agricultural Research. In: Persley, G.J. and Lantin, M.M. (eds) *Agricultural Biotechnology and the Poor: Proceedings of an International Conference, Washington, DC, 21–22 October 1999.* Consultative Group on International Agricultural Research, Washington, DC pp. 118–120.

Doyle, J.J. and Persley, G.J. (1996) *Enabling the Safe Use of Biotechnology: Principles and Practice.* Environmentally Sustainable Development Studies and Monograph Series 10, World Bank, Washington, DC.

Gopo, J. (2000) Zimbabwe. In: Tzotzos, G.T. and Skryabin, K.G. (eds) *Biotechnology in the Developing World and Countries in Economic Transition.* CAB International, Wallingford, UK, pp. 187–193.

Lele, U., Lesser, W. and Horstkotte-Wesseler, G. (2000) *Intellectual Property Rights in Agriculture: the World Bank's Role in Assisting Borrower and Member Countries.* Environmentally and Socially Sustainable Development Series, World Bank, Washington, DC, 87 pp.

Persley, G.J. (1990a) *Beyond Mendel's Garden: Biotechnology in the Service of World Agriculture.* CAB International, Wallingford, UK, 155 pp.

Persley, G.J. (ed.) (1990b) *Agricultural Biotechnology: Opportunities for International Development.* CAB International, Wallingford, UK, 495 pp.

Persley, G.J. and Lantin, M.M. (2000) *Agricultural Biotechnology and the Poor: Proceedings of an International Conference, Washington, DC, 21–22 October 1999.* Consultative Group on International Agricultural Research, Washington, DC.

Woodend, J.J. (1995) *Biotechnology and Sustainable Crop Production in Zimbabwe.* OECD Technical Papers No. 107, Paris, 79 pp.

# Latin America

# Brazil

**11**

## Marie Jose Amstalden Sampaio

| | |
|---|---|
| Area ($km^2$) | 8.511m |
| Cropland | 1% |
| Irrigated cropland | 28,000 $km^2$ |
| Permanent pasture | 22% |
| Population | 171.8m |
| Population/$km^2$ | 20 |
| Annual population growth rate | 1.6% |
| Life expectancy (men) | 59 yrs |
| (women) | 69 yrs |
| Adult literacy | 83% |
| GDP (1998) | US$1.03 tr |
| GDP per head | US$6100 |

| | |
|---|---|
| Annual growth in real GDP (1998) | 0.5% |
| Agriculture as % of GDP | 14.8% |
| Value of agricultural exports | US$2.8 bn |
| Agricultural products as % of total exports | 8.8% |

Major export commodities: coffee, soybean

Major subsistence commodities: maize, soybean, rice, sugar cane, cocoa, citrus

## Summary

Brazil is the world's fifth largest country. It is still heavily dependent on agricultural products. Most of the agricultural biotechnology research and development is supported by government (heavily financed by the World Bank) within government institutions. There is an urgency to stimulate the participation of the private sector at early stages of R & D of potential agricultural biotechnology products.

Joint ventures have been one of the most effective means of transferring technology, production and marketing know-how. A need exists for bioproducts and processes, and regulatory agencies are needed to guarantee long-term business investment in agricultural biotechnology.

## Introduction

The needs and expectations from biotechnology are great. Plant biotechnology is required to generate the knowledge to produce new plants with a higher yield capacity and with better stress resistance. Some also expect the technology to produce plants that can be cultivated with lower inputs of environmentally toxic chemicals, plants that have additional value for specific niche markets, plants that can be turned into biofactories, plants that can better harvest and transform sunlight and plants that will be more resistant to ultraviolet (UV) radiation (an effect of the diminishing ozone layer). The research and development (R & D) agenda is extensive.

Agriculture productivity will have to be boosted by the introduction of new plants, even wild species, to produce enough food for the world population, expected to reach around 8 billion by 2030 (UN Population Division – Department of Economic and Social Affairs, 1998). The time frame is short, but with biotechnological tools these traits will not sound so Utopian in 10 years (van Montagu, 1998).

Although humans have cross-pollinated plants and cross-bred animals for centuries to suit their needs, recent technological advances in molecular biology have provoked reactions from different parts of society, ranging from optimism to cautiousness to moral outrage (Background, *Sustainable Development* 30 (1), September 1999). Throughout the world biotechnology managers are involved in discussing the pros and cons with the press, politicians, policymakers, consumer representatives and non-governmental organizations (NGOs). The dialogue must improve in the scientific, social, cultural and ethical areas to resolve uncertainties and identify areas of consensus (Macedo Oda *et al.*, 2000).

## Opportunities and Constraints

As highlighted in the on-line *Nature* supplement *Science in Latin America* (www.nature.com/server-java/Propub/nature/398A001A0.frameset context=search), the region enjoys a unique opportunity to win a more prominent place in the world of science. In Brazil, many lines of research and development are already benefiting from the application of biotechnology tools such as marker-assisted plant and animal breeding, genomic mapping of several species, embryo transfer applied to different animal species, genetic resources characterization and conservation and transgenic products (see Box 11.1).

The same *Nature* review article has identified, among others, three difficulties that relate to this forum: the lack of regional integration in science, scientists' reluctant acceptance of the free market and a failure to acknowledge the importance of intellectual property rights (IPR) in modern research. Biotechnology applications are teaching new lessons and adding new challenges in all three aspects.

Recognizing IPR is a behavioural change that will come as a consequence of understanding the system. Solutions, however, must accompany this acceptance. It is already far from easy to develop transgenic products. It is extremely difficult and expensive to negotiate licence agreements. For example, nine different companies are involved in a project to commercialize a papaya cultivar that carries resistance to a virus disease. Alliances and joint projects with the international agricultural research centres (IARCs), US universities and other centres of excellence within the region could add strength to negotiations.

The integration of markets has made genetically modified (GM) seeds and GM processed food hit Brazil faster than the internal research organization could deploy them. Consumers are in a confusing situation, because they receive no warning and are advised by conflicting information in the press and on the internet. Scientists are only beginning to learn how to deal with the

**Box 11.1.** Transgenic plants – some examples from Brazil.

*Brazilian maize to produce growth hormone*

Developed by the Molecular Biology and Genetic Engineering Centre of the State University of Campinas (Unicamp) and the Chemistry Institute of the University of São Paulo (USP), these plants are ready to produce 250 g of the hormone per ton of seeds – enough to treat hundreds of patients for months. The hormone is identical to the human form and therefore better than the bacterial source, which has one extra amino acid. It proved to be cheaper to produce and extract.

*Papaya resistant to Brazilian strain of ringspot virus*

Developed in collaboration with Cornell University, these plants have been tested in greenhouses in Geneva, New York, and have now been transferred to Embrapa in Brasilia for field tests. In 2 years they should be ready for large-scale tests and should be as successful as their cousins being planted in Hawaii. The technology will bring the opportunity of papaya cultivation back to small farmers in areas where the crop has been decimated by virus disease. However, if the antibiotic marker is proved to be a real problem under Brazilian conditions, then another 4–5 years will be necessary to reconstruct the material.

*Common beans resistant to golden mosaic virus*

Developed by Embrapa (Rice and Beans Centre), these plants are undergoing greenhouse tests after a long research period, due to the difficulty of adapting existing technology to the specific virus strain. Researchers expect to complete the cross-breeding of the characteristic into commercial lines in 2–3 years.

constant questions about the safety of their work. The fact is that, with the exception of well-known traits already tested in the USA and consumed by millions in the late 1990s (e.g. the 'Roundup-Ready' (RR)-soybean), more research is needed to clarify basic questions in different environments. Tropical agriculture is very different from the temperate fields where most products have been tested. Protocols are required for field trials, risk assessment for environmental and food safety, registration of products and public acceptance. The need is urgent, because these are constraints that will intensify as GM organisms (GMOs) become an integral part of the research agenda in the region.

## Challenges

Brazil still depends heavily on agriculture and a continuing supply of new technologies to increase its competitive advantage in the region and in export markets. Increased exports mean increased benefits to the general population. Poverty reduction programmes are always dependent on how well the country can manage its economy, including support for R & D.

Agricultural biotechnology promises to increase yields and market value for farmers. It promises to produce plants that will grow in harsh environments with less need for chemical input, therefore protecting the environment, to produce new cultivars with increased nutritional composition and to reduce postharvest storage losses. The greatest research challenge, and maybe one that has not yet been seriously tackled by most of those holding the necessary knowledge, is the transfer of these new characteristics to social crops, to staple crops that will feed the hungry populations. We also need to simplify the use of traits, making them available to the small-scale farmers in developing countries. The Rice Biotechnology Programme, financed by the Rockefeller Foundation, is an excellent example of this approach (Conway, 1999).

## Intellectual Property Rights

The IPR challenge is directly linked to the application of biotechnology tools outside the corporate world, where companies can afford to acquire rights, make alliances or develop innovations on their own. Because the patent system has undergone a process of regulatory globalization and harmonization and Trade-Related Aspects of Intellectual Property Rights (TRIPs) have obliged most developing countries to move to some level of recognition of IPR in agriculture, problems that were not common to research managers regarding IPR are now causing concern. The scope of patentable subject-matter has also been given an inclusive interpretation and restrictions on patentability have been narrowly interpreted, enabling applicants for biotechnology patents to overcome existing bars (Drahos, 1999).

Most of the basic tools used in many biotechnology projects in developing countries (promoters, markers, transformation processes (biolistics, *Agrobacterium*), broad-scope enabling techniques) have been patented by their inventors in industrial countries and are in the hands of a few large life-sciences companies. Some of these R & D projects in developing countries are nearing completion. Initial material transfer agreements (MTAs) covered only research applications and laboratories, and some companies are now facing difficult negotiations to allow licensing of the right of commercialization of their transgenic products (Cohen, 2000).

## Regulatory Issues

The regulatory/risk-assessment challenge encompasses: (i) food and environmental safety concerns, which can exist anyway when dealing with a new technology; (ii) financing of these extra phases of research; (iii) ethical and religious concerns; (iv) public awareness; (v) right of choice by consumers; (vi) adequate labelling; and (vii) the fact that genetic engineering has turned into a hot political issue for opposition groups, who attack globalization, competition markets, technological substitution, monopolies/oligopolies of knowledge and of seeds by transnationals and other concerns.

Despite efforts in Brazil since 1995 to develop biosafety legislation and to establish a regulatory infrastructure to deal with the arrival of transgenic crops in the market in an organized way, there is still a battle over soybean.

The commercial introduction of Monsanto's RR-soybean has coincided with strong European Union (EU) refusal of transgenic foods since late 1998 and with the recent (1998–2000) and aggressive acquisition of commodity seed companies, operated with national capital, by the same transnational companies that are being accused of building a potential global monopoly in agricultural biotechnology. The parallel approach of European supermarket chains, with promises of premium prices for GMO-free soybean of certified origin, has inflamed local politicians and farmers, who were looking for new export dollars. This has also given opposition groups a special tool to fight against the technology and against Monsanto and other biotechnology companies.

A critical point in the growing confusion was reached when Greenpeace and the Brazilian consumer's institute (IDEC) filed an injunction against Monsanto and against the National Biosafety Committee (CTNBio). They asked a judge to invalidate the approval for commercialization already given by CTNBio, because RR-soybean could be harmful to the environment and because more tests were needed. Higher courts were to review the appeal case. News reports in 1999–2000 suggested that more than 2 million ha of RR-soybean were planted with illegal seeds brought from Argentina, possibly resulting in the appearance of new diseases not common in Brazilian fields. Five thousand identification test kits were acquired by the state government of Rio

Grande do Sul to guarantee, for commercial reasons, that the state is a GMO-free zone.

Public opinion is not well informed, with some media publishing inaccurate comments. Can this situation be corrected? The answer is yes, but players at all levels must help. Scientists must enter the public dialogue instead of debating among themselves in scientific journals (Burke, 1999; Ewen and Pusztai, 1999; Horton, 1999; Kearns and Mayers, 1999; Losey *et al.*, 1999; Millstone *et al.*, 1999). Dissemination of information based on trusted sources must be maximized for the benefit of society, showing potential benefits, potential risks and what is being done to increase knowledge in these areas.

## Conclusions

Apart from well-trained scientists, two items are always part of the recipe for a successful research project: funds and tools, both tangible and intangible, such as IPR. We must now educate our politicians and the public and involve lawyers in all future agreements involving research in biotechnology.

We must be careful not to infringe the rights of others when developing new biotechnological projects in developing countries, where minimum TRIPs regulations are now in place. This also applies when a new transgenic plant or animal is going from the laboratory to the market-place. Depending on the case, a complete inventory of the 'freedom to operate' might be complicated and costly and has not been in the list of concerns of scientists until now. There are, of course, genes in the public domain but most of the well-characterized traits and processes are patented in industrial countries and with TRIPs in place there might be a chance that the patents would stand in developing countries. Access to this information is urgent.

IARCs could develop, for the benefit of developing countries, more comprehensive partnerships with the private sector, with US universities and other advanced research institutions (Serageldin, 1999). This would give developing countries access to a minimum intellectual property (IP) platform that would guarantee that new products developed by their research institutes would reach farmers and consumers.

Another option would be to validate new discoveries, new methodologies (e.g. those used to modify salinity resistance or control fruit maturation) and negotiate non-exclusive licences for different applications, with regional market segmentation.

An interesting action that has been suggested many times would be for the IARCs to act on the training of researchers, not only in biotechnology skills but also on IP management and policy development. Challenges were identified and options for solutions were proposed at a recent regional meeting in Costa Rica (September 1999) (Cohen, 2000). Some suggestions (Sampaio, 1998) included:

- Development of a national competence in IPR, through improved training.
- Dissemination of IPR information and procedures through workshops, short courses and seminars.
- Dissemination of knowledge and use of IPR systems as an important tool for technological development.
- Training in negotiation skills, MTAs and contract design (case-studies).
- Provision by the Consultative Group on International Agricultural Research (CGIAR) of legal support on the licensing and use of proprietary technologies and on contract management.
- Financing of in-house development of biotechnological tools to enhance bargaining power when accessing IP in the private sector.
- Provision by donors and the IARCs of licences for enabling technologies, acquired from the owners in the private or academic sectors in developed countries.

Helping developing countries resolve the biosafety and risk assessment issues is a major task for IARCs and development agencies. Much more detailed research will be needed to change the present lack of public acceptance. This seems to be a fine example for regional collaboration. IARCs could use their credibility, choose case-studies and issue a detailed manual to guide national agricultural research organizations. This could also cover use of data to inform the public to ease their concerns about the use of new technologies for genetic improvement of crops and livestock.

## References

Burke, D. (1999) No GM conspiracy. *Nature* 401, 640.

Cohen, J.I. (2000) Managing intellectual property – challenges and responses for agricultural research institutes. In: Persley, G.J. and Lantin, M.M. (eds) *Agricultural Biotechnology and the Poor: Proceedings of an International Conference, Washington, DC, 21–22 October 1999*. Consultative Group on International Agricultural Research, Washington, DC, pp. 209–217.

Conway, G. (1999) *The Rockefeller Foundation and Plant Biotechnology*. 24 June. www.biotech-info.net/gordon_conway.hyml).

Drahos, P. (1999) Biotechnology patents, markets and morality. *European Intellectual Property Review* 21 (9), 441–449.

Ewen, S.W.B. and Pusztai, A. (1999) Effects of diets containing genetically modified potatoes expressing *G. nivalis* lectin on rat small intestine. *Lancet* 354, (9187) (www.thelancet.com).

Horton, R. (1999) Genetically modified food: 'absurd' concern or welcome dialogue? *Lancet* 354 (9187) (www.thelancet.com).

Kearns, P. and Mayers, P. (1999) Substantial equivalence is a useful tool. *Nature* 401, 640.

Losey, J.E., Rayor, L.S. and Carter, M.E. (1999) Transgenic pollen harms monarch larvae. *Nature* 399, 214.

Macedo Oda, L., Correa Soares, B.E. and Valadares-Inglis, M.C. (2000) Brazil. In:

Tzotzos, G.T. and Skryabin, K.G. (eds) *Biotechnology in the Developing World and Countries in Economic Transition.* CAB International, Wallingford, UK, pp. 43–52.

Millstone, E., Brunner, E. and Mayer, S. (1999) Beyond 'substantial equivalence'. *Nature* 401, 525–526.

Sampaio, M.J. (1998) *Intellectual Property Rights: an Important Issue for the Brazilian Agriculture R & D and Related Agribusiness.* World Bank Meeting on Biotechnology and Biosafety, Washington, DC.

Serageldin, I. (1999) Biotechnology and food security in the 21st century. *Science* 285, 387–389.

van Montagu, M.V. (1998) How and when will plant biotechnology help? In: Waterlow, J.C. *et al.* (eds) *Feeding a World Population of More Than Eight Billion People.* Oxford University Press, Oxford.

# Colombia

## Ricardo Torres and Cesar Falconi

| | | | |
|---|---|---|---|
| Area ($km^2$) | 1.139m | GDP per head | US$6600 |
| Cropland | 4% | | |
| Irrigated cropland | 5300 $km^2$ | Growth in real GDP (1998) | 0.2% |
| Permanent pasture | 39% | | |
| | | Agriculture as % of GDP | 19% |
| Population | | | |
| (1999 est.) | 39.3m | Valuc of agricultural | |
| Pop. per $km^2$ | 34 | exports | US$1.3 bn |
| Ann. pop. growth rate | | Agricultural products as | |
| (1999 est.) | 1.85% | % of total exports | 16% |
| Life expectancy (men) | 66.5 yrs | Major export commodities: | |
| (women) | 74.5 yrs | coffee, flowers, banana | |
| Adult literacy | 91.3% | Major subsistence commodities: | |
| | | bananas, rice, maize, sugar cane, | |
| GDP (1998 est.) | US$254 bn | cocoa, beans, oil-seed, vegetables | |

## Summary

Colombian agriculture has been historically highly protected, with considerable state intervention. This has been successful in creating food self-sufficiency, but has had a negative impact on Colombia's international competitive position.

Coffee continues to be Colombia's leading export cash crop, although its relative importance has declined in recent years due to growth in other agricultural exports. Cattle remain important on large and small farms.

National efforts in biotechnology aim to create a new capability, ensuring that knowledge and techniques are acquired and developed locally. By selective promotion and application to developments that will improve the comparative advantage of Colombian products, the Colombian government hopes to ensure a wider distribution of benefits – particularly in rural areas – of technical change. At the same time there is concern to maintain the ongoing health and productivity of the environment.

## Introduction

Biotechnology's impact on agricultural production in the medium and long term in developing countries requires urgent discussion. If developing countries wish to participate in the global biotechnology revolution, national research organizations are challenged to incorporate the basic capabilities and tools of modern biotechnology into productive processes. This is particularly important for agriculture, which is often the largest economic sector in developing countries. Agriculture plays a vital role in income and employment generation as well as foreign-exchange earnings. Biotechnology could effectively support developing countries by fostering sustainable development and creating and maintaining competitive positions in international agricultural markets.

Physical, human and financial resources available for agricultural investments are one indicator of efforts to strengthen or create these capabilities. In addition, information on the size, structure and content of public research is needed to improve policy decisions, clarify the roles of the public and private sectors and support public-sector implementation of biotechnology research.

The International Biotechnology Service (IBS) of the International Service for National Agricultural Research (ISNAR) organized a survey of research indicators on agricultural biotechnology research to determine how relevant resources are mobilized and used to implement such research. The survey sought to analyse and report on policy implications of the data collected, conduct policy analyses and disseminate data to relevant national decision-makers. The survey also aimed to strengthen capacity to compile research indicators and to identify gaps and needs to implement biotechnology programmes and mobilize the required resources.

Descriptive information is provided on biotechnology programmes and institutions in Colombia. Information is also provided on physical, human and financial resources available for agricultural biotechnology, and is also included on some of the research projects being carried out in agricultural biotechnology in Colombia. Detailed statistical and institutional information, collected through the survey on research indicators for agricultural biotechnology research and on the development and current situation of the most relevant public and non-public organizations involved in agricultural biotechnology research activities in Colombia, is provided in ISNAR Discussion Paper No. 5, on which much of this report is based (see Torres and Falconi, 2000).

## Evolution of Agricultural Biotechnology

Biotechnology was first used in Colombia in 1973, when the National Coffee Research Centre (CENICAFE) applied tissue-culture techniques to complement its research on industrial aspects of coffee production and processing (Sasson, 1993). Flower export companies also used tissue-culture techniques to improve their competitiveness in the international market. Americaflor was the first company to do this, in 1978. Around this time, the first tissue-culture laboratories at universities were created at the National University in Bogota and Caldas and Javeriana universities. In 1980, the Colombian Institute for Science and Technology (COLCIENCIAS) supported the establishment of the Colombian Association of Tissue Culture, which is composed of public and private laboratories.

The Biotechnology Group of the National University carried out reviews of biotechnology at the national level in 1983 and 1986. Through this diagnostic study, the main institutions involved in biotechnology were identified and the main areas of biotechnology research were defined. This process ended with the creation of the Institute of Biotechnology of the Universidad Nacional de Colombia (IBUN) in 1987.

In 1985, the biotechnology unit at the International Centre for Tropical Agriculture (CIAT) was established to conduct research on cassava, beans, rice and forage, using tissue-culture techniques. This unit started to use genetic engineering techniques in 1990. They maintain close collaboration with many Colombian research organizations through biotechnology projects and training.

The Colombian Agricultural Institute (ICA), now known as CORPOICA (Agricultural Research Colombian Corporation), initiated biotechnology research in 1989. In 1993, the Biotechnology Programme at CORPOICA was established, after biotechnology was recognized as an essential tool for the development of agricultural research. CORPOICA made special efforts to set priorities in their biotechnology research at the national level, with the participation of farmers and researchers.

COLCIENCIAS created the National Biotechnology Programme in November 1991. This is one of the 11 programmes under COLCIENCIAS. This programme emerged from the belief that biotechnology was a fundamental factor in the incorporation of leading-edge technologies in production (Torres, 1993). The programme succeeded in obtaining fiscal incentives for biotechnology investments, which can be deducted from net income for tax purposes (Hodson and Aramendis, 1995; Sanint, 1995). The National Biotechnology

Programme developed the first national strategy for biotechnology development (COLCIENCIAS, 1993), consisting of the following elements:

- Training of qualified researchers.
- Consolidation of the Colombian biotechnology community.
- Strengthening of links between production and research.
- Formulation of legislation for intellectual property rights, biosafety and the use of germ-plasm.
- Capacity to monitor scientific and industrial biotechnology activities.

The COLCIENCIAS (1993) document was the first recognition by the Colombian government that biotechnology for development is important. The strategy was general, however, and its implementation is still ongoing. COLCIENCIAS allocated US$500,000 to the Biotechnology Programme to fund agricultural biotechnology projects and to implement the strategy in 1991. During 1992–1999, COLCIENCIAS provided an annual average of about US$1 million to agricultural biotechnology projects (COLCIENCIAS, 1999).

COLCIENCIAS has the responsibility for conducting a strategic planning exercise and the main input is to set priorities for biotechnology research in agriculture, medicine and the environment. A clear policy and national strategy for the development of biotechnology are still lacking. The Biotechnology Programme has, however, recently published the Biotechnology Strategic Plan for 1999–2004, after analysing biotechnology in Colombia (COLCIENCIAS, 1999). The objectives of the strategic plan are to encourage:

- Formulation of policies to promote national, international, public and private investment through tax incentives.
- Technology transfer, adoption, distribution and commercialization of biotechnological products by interested private entities.
- Biotechnology firms to use a proper selection of techniques and ensure the quality of final products and/or processes.
- Support for training at all levels.
- Universities and research centres to establish linkages and partnerships with the private sector to transfer biotechnology.
- Market niche identification within the country and in the international market-place.

According to the plan, some agricultural biotechnology research areas should be strengthened, such as genomics and the application of biotechnology for breeding, biopesticides and biofertilizers.

Colombia made good progress by establishing the necessary regulatory framework for agricultural biotechnology in 1994.

### Intellectual property rights

Colombia, as a member country of the Andean Pact, approved in 1993

Resolution 344, which is the Regime for Industrial Property. Colombia regulated the Resolution in 1994 by Decree 117. The Resolution protects any invention (including biotechnology products or processes) if they are novel inventions and have an industrial application. The resolution, however, emphasizes that discoveries are not inventions. Likewise, it specifies that species and animal breeds and essentially biological processes to obtain them will not be patentable. It contains no stipulation on plants and plant varieties, so it could be interpreted that, if they comply with the three requisites of novelty, inventive step and industrial application, they are then patentable (Salazar, 2001). Commerce and Industrial Superintendence, the agency in charge of implementing the Regime, has received about 200 patent requests for biotechnological products and processes since 1991. Only ten of them have been approved (Rueda, 1996).

Plant varieties are also protected by Common Resolution 345 of the Andean Pact, which contains common rules for plant protection. Colombia regulated the Resolution by Decree 533 in 1994 and ICA Resolution 1853 in 1995. Colombia has been a member of the Union Internationale pour la Protection des Obtentions Végétales (UPOV) since 1978.

## Biosafety regulations

ICA established Resolution 3492 in December 1998 to regulate the introduction, production, release and commercialization of genetically modified organisms. ICA is the only organization to regulate and control transgenic products in the country. The Agricultural Biosafety Committee was set up in 1998 as well, with ten representatives from government and leading national biotechnology research institutes (De Kathen, 1999).

These regulations will hopefully encourage private-sector involvement in agricultural biotechnology research in Colombia. There are some private commercial groups in agricultural biotechnology. Colombia Biotechnology Company (ECB) specializes in mass propagation of fruit and was established in 1990. Green Seed Ltda. (GSL), a new private company, applies tissue culture to produce ornamentals. The Colombian Tobacco Company initiated its biotechnology applications in 1988, as did the Colombian Sugarcane Research Centre (CENICAÑA) in 1993.

Torres (1993) showed that the number of organizations applying biotechnology techniques doubled in 1993, focusing on tissue culture. They have now started to use molecular markers and genetic engineering. Hodson and Aramendis (1998) indicated that there were about 47 entities using biotechnology techniques in 1998 (26 were private). Tissue culture was still the main technique used, but several groups said they were using genetic transformation, molecular markers, diagnostic techniques and microbiology. The studies showed that the lack of qualified researchers is one of the main constraints to the development of biotechnology in Colombia.

There has been a significant growth of private laboratories closely tied to export commodities, such as flowers and bananas. The more important private laboratories are Americaflor, GSL and ECB. The first two have recently been taken over by multinationals (Dole and Norvartis) to develop vertical control and integration of crops. CORPOGEN mainly conducts molecular biology to produce and provide inputs for other laboratories and CETELCA applies biotechnological techniques to animal breeding.

Other recent important developments include the expansion of the laboratories of the Universidad Javeriana, the use of bioreactors by the Universidad Catolica del Oriente, the establishment of new molecular biology at the Von Humboldt and Sinchi institutes and the creation of germ-plasm banks at CORPOICA.

The government of Colombia has placed priority on the application of biotechnology for the development of agriculture since the early 1990s. Even with the recent COLCIENCIAS Biotechnology Strategic Plan, there is still a lack of clear policy and national priorities on biotechnology. There is a need to share biotechnological resources among the biotechnology research groups. Relationships with international advanced research centres need to be strengthened.

## Current System

The most relevant public and private organizations currently involved in agricultural biotechnology research in Colombia are listed in Table 12.1. This sample of surveyed organizations represents more than 70% of the total expenditures on agricultural biotechnology research in the country.

Three entities not included in the survey deserve mention. The Centre for Biological Research (CIB) is using molecular markers for the production of biopesticides; Universidad Catolica del Oriente has gained experience in applying tissue-culture techniques for trees and banana mass propagation.

The government of the Netherlands, through its Directorate General for International Cooperation, stimulates biotechnology development in selected developing countries through the implementation of an international cooperation programme. The Netherlands government has allocated US$3.8 million for the establishment of the Colombian Agricultural Biotechnology Programme (ABP) (Spijkers, 1998). ABP started in 1993, but it took 4 years to formalize the Netherlands government's support in a contractual arrangement. The Centre for Agricultural and Livestock Studies (CEGA), a non-governmental organization (NGO) that carries out socio-economic studies on agriculture, was selected to manage the programme. The main objective of the programme is to contribute to the improvement of the socio-economic and productive conditions of small farmers of the Caribbean coast, through the applications of biotechnology.

The programme applies a participatory bottom-up approach, which means that small farmers (users of biotechnology outputs) participate in defining, prioritizing, monitoring and evaluating activities and projects. The programme is directed by an executive committee composed of representatives of small farmers, government (Planning Department, Ministry of Agriculture and COLCIENCIAS) and research and academic sectors (CORPOICA and IBUN). The committee decided that 85% of resources would be orientated to obtaining virus-free plantlets of plantain, cassava and *ñame* and the remaining 15% to research on the development of transgenic plants. The main executors of the research projects are CORPOICA, the National University of Colombia, CIAT and regional universities.

A list of all institutions engaged in agricultural biotechnology in Colombia is presented in Torres and Falconi (2000). The list includes a brief outline of ongoing activities in biotechnology. Torres and Falconi (2000) also include in their report the organizational structure of the main agricultural research groups.

## Opportunities

The survey results of the ten institutions that responded to our questionnaire (Table 12.1) provides the basis for the following section. The detailed statistical tables from the original survey are included in Torres and Falconi (2000).

### Overview

The number of researchers in the agricultural biotechnology research system in Colombia during 1985–1997 has grown much more than the research expenditures. This has led to an ongoing decline in the expenditures per researcher (in real international dollars from $60,600 in 1985 to $27,200 in 1997). It is expected, however, that this trend could be reversed with the recent contribution of the Netherlands government to the development of biotechnology in Colombia.

Although the research intensity ratio has grown annually, the percentage of the agricultural biotechnology research expenditures in relation to the agricultural GDP is minimal – less than 0.01% on average. The percentage of agricultural biotechnology research expenditures to total agricultural research expenditures was around 2.0, on average. There is no set rule about how much of the agricultural research budget should be allocated to biotechnology. For comparison, however, the Consultative Group on International Agricultural Research (CGIAR) spent about 8% of its budget on biotechnology research in 1997 and the USA allocated 13% of its agricultural research expenditures to biotechnology in 1992 (Caswell *et al.*, 1994; Fuglie *et al.*, 1996).

**Table 12.1.** Overview of Current Agricultural Biotechnology Research Institutes, 1997 (from ISNAR–IBS Survey (Torres and Falconi, 2000)).

| Institutional category | Executing agency | Biotechnology research focus | Laboratories | No. researchers | Started biotechnology research |
|---|---|---|---|---|---|
| Public enterprise | Agricultural Research Colombian Corporation (CORPOICA) | Genetic engineering<br>Molecular markers<br>Tissue culture<br>Crops and animals | 9 | 45 | 1993 |
| University | Biogenesis<br>Universidad de Antioquia<br>Faculty of Medicine | Molecular markers<br>Tissue culture<br>Cattle | 3 | 20 | 1990 |
| | Institute of Biotechnology<br>Universidad Nacional de Colombia (IBUN) | Genetic engineering<br>Molecular markers<br>Tissue culture<br>Crops | 9 | 24 | 1987 |
| | Plant Biotechnology Unit<br>Pontificia Universidad Javeriana (PUJ) | Genetic engineering<br>Molecular markers<br>Tissue culture<br>Fruits | 5 | 12 | 1985 |
| Private non-commercial | National Coffee Research Centre (CENICAFE) | Genetic engineering<br>Molecular markers<br>Tissue culture<br>Coffee | 1 | 9 | 1978 |

| | | | | | |
|---|---|---|---|---|---|
| | Colombian Sugarcane Research Centre (CENICAÑA) | Molecular markers<br>Tissue culture<br>Sugar cane | 3 | 3 | 1993 |
| Private Commercial | Colombian Biotechnology Company (ECB) | Tissue culture<br>Banana, flowers, papaya | 2 | 2 | 1990* |
| | Green Seed Ltda (GSL) | Tissue culture<br>Banana, sugar cane, trees | 1 | 5 | 1995 |
| | Americaflor | Tissue culture<br>Ornamentals | 2 | 5 | 1978† |
| | Colombian Tobacco Company (COLTABACO) | Tissue culture<br>Tobacco | 3 | 6 | 1988 |
| Total | | | 38 | 131 | |

*In 1997 ECB was constituted after taking over Meristemos SA (private company).

†In 1979 the meristem laboratory was established and in 1993 biotechnology research was intensified.

## Human resources

The number of researchers increased more than 20 times between 1985 and 1997, and researchers holding a doctorate degree increased almost nine times. This group of professionals will be the basis for the development of agricultural biotechnology advances in the future. It is important that about 55% of the researchers in the private sector held a postgraduate degree in 1997. Around 77% of the researchers are located in only four research organizations, which belong to the public sector: CORPOICA, Biogenesis, IBUN and Pontificia Universidad Javeriana (PUJ).

Considering the extent of donor intervention and international scientific interaction to enhance the development of agricultural biotechnology, only one expatriate researcher was identified in the survey. This indicates low links to the international scientific community. This is surprising because Colombia has established new postgraduate programmes during the last 15 years and extensive international cooperation.

During 1985–1997, about 47% of the research staff were female, with 28% located in the public research enterprise, 49% at universities, 8% in the private non-commercial sector and 15% in the private commercial sector.

Detailed information on the personnel structure of each biotechnology programme is presented in Table 12.2. Downer *et al.* (1990) suggested a minimum efficient size for research groups in agricultural biotechnology. For genetic engineering and tissue culture, they recommended a ratio of one researcher for two support personnel (technicians). The number of technical support staff per researcher is less than one for almost all surveyed research organizations, except for ECB and Americaflor (private companies). The number of other support staff (administrative, secretarial and labourers) per

**Table 12.2.** Staffing structure in agricultural biotechnology, 1997 (from ISNAR–IBS Survey (Torres and Falconi, 2000)).

| | Professional | | | Support | | |
|---|---|---|---|---|---|---|
| Staff category | Management | Researchers | Total | Technical | Other | Total |
| CORPOICA | 5 | 40 | 45 | 13 | 32 | 45 |
| Biogenesis | 3 | 17 | 20 | 3 | 2 | 5 |
| IBUN | 5 | 19 | 24 | 4 | 9 | 13 |
| PUJ | 1 | 11 | 12 | 21 | 2 | 23 |
| CENICAFE | 1 | 8 | 9 | 3 | 0 | 3 |
| CENICAÑA | 1 | 2 | 3 | 1 | 0 | 1 |
| ECB | 1 | 1 | 2 | 1 | 1 | 2 |
| GSL | 2 | 3 | 5 | 2 | 6 | 8 |
| Americaflor | 1 | 4 | 5 | 4 | 2 | 6 |
| COLTABACO | 3 | 3 | 6 | 2 | 3 | 5 |
| Total | 23 | 108 | 131 | 54 | 57 | 111 |

researcher is less than one in most public and private research organizations, except for ECB and the Colombian Tobacco Company (COLTABACO) (private companies). In general, most of the research organizations show a low ratio of technical support to researchers.

## Financial resources

In agricultural biotechnology research expenditures in Colombia, public research accounted for almost 34% of the total during 1985–1997. The participation of public universities is about 27% and the private sector (commercial and non-commercial) 39%. Public- and private-sector expenditures are growing. CORPOICA had the highest annual growth rate (around 63%), whereas private non-commercial groups showed the lowest annual growth (9%).

Personnel expenditures as a percentage of total expenditures were quite high for Biogenesis and IBUN (public universities). The cost structures for CORPOICA and PUJ looked more normal in 1997. Most of the private commercial and non-commercial entities showed lower personnel cost than the public sector, which means that private-sector researchers have more operational expenditures available to conduct their research.

The sources of funding of agricultural biotechnology research during 1985–1997 show that the government share in total expenditures has been considerable for the public research enterprise and the biotechnology programmes at the universities. Government contribution represents about 35%, on average, of the total agricultural biotechnology expenditures. In 1997, CORPOICA, Biogenesis and IBUN's expenditures were 70% funded by the government. During the period of analysis, the donor share in total agricultural biotechnology research expenditures amounted to almost 16%. About 70% of PUJ's budget was funded by donor contributions in 1997. The other research organizations showed nil or minimal funding from donors.

Some public universities (Biogenesis and IBUN) are funding their biotechnology research activities from non-traditional sources, such as sales of products and services. Although these sources of funding are minimal, they have increased since 1997.

The biotechnology activities carried out by private, non-commercial organizations are largely funded by levies on the sales of their products (e.g. coffee and processed sugar). Private commercial organizations, on the other hand, are fully financed by the sales of their products.

Donor share of total expenditures is almost 16%. Most of the donor contributions went to infrastructure and operational costs during the period of analysis, with these two categories accounting for almost 90%. The share of operations to total donor contributions has slightly decreased, while infrastructure increased. This may be explained by the sharp increase in numbers of researchers and the need to update their laboratories.

## Research Focus

In 1997, 60% of the researchers involved in agricultural biotechnology research in the main ten Colombian organizations listed in Table 12.2 focused on crops research. CORPOICA and Biogenesis are the only two research organizations that are applying biotechnology in livestock research. About 40% of the researchers are dedicated to livestock biotechnology research, reflecting the contribution of livestock to the value of agricultural production in Colombia, which was 40% in 1997.

The emphasis of the biotechnological techniques used by researchers is noteworthy. We used the gradient of biotechnologies proposed by Jones (1990), where traditional plant tissue culture is at one extreme requiring less scientific knowledge, time and financial resources. At the other extreme is genetic engineering in plants and animals. The distribution of researchers on the gradient is an indicator of the biotechnology development of a country. In Colombia, about 35% of the researchers are utilizing advanced techniques, including genetic engineering for plants, embryo manipulation for bovines, molecular markers for plants, animals and microorganisms, massive propagation of *in vitro* plants through bioreactors and fermentation processes to produce biopesticides. The remaining 65% are using less sophisticated techniques such as tissue culture and nitrogen fixation for crop improvement.

CORPOICA, Biogenesis, CENICAFE and IBUN apply sophisticated biotechnology techniques and employ highly qualified scientists. CORPOICA and IBUN have built a good infrastructure and a highly skilled staff. CENICAFE has the best plant genetic engineering research team and Biogenesis has developed strong expertise in the use of molecular markers.

CENICAÑA and PUJ have achieved an intermediate level of development. CENICAÑA recently established its biotechnology unit to apply genetic engineering and molecular markers in sugar-cane breeding. PUJ has considerable experience in the application of tissue-culture techniques and recently started to use genetic engineering for local fruit. The other organizations in Colombia are still in the early stages of developing biotechnology research.

In most developing countries, private-sector organizations use mainly less sophisticated techniques, such as tissue culture, biofertilizers and bioinsecticides. These are less costly, less risky and closer to the market. Another factor is the limited enforcement of intellectual property rights. About 77% of the private-sector researchers are applying less sophisticated techniques. The more advanced research techniques, which are more expensive and where the pay-offs are more uncertain, are used mainly by public-sector organizations, involving 41% of public researchers. A significant proportion of public researchers (59%), however, use less advanced techniques.

The division of labour implied by these results between the public and private sector should be taken into account by research leaders and decision-makers in the allocation of resources for the development of biotechnology, to promote partnerships between both sectors.

## Conclusions

The ISNAR survey of the most relevant public and non-public organizations involved in agricultural biotechnology research in Colombia yielded the following:

- CORPOICA, Biogenesis, CENICAFE and IBUN are using sophisticated biotechnology tools and have highly qualified scientists. The other organizations in Colombia are still in the early stages of developing biotechnology research. Most of the organizations started or intensified their biotechnology research activities in the early 1990s.
- Partnerships with some international advanced research centres and universities (such as CIAT, Cornell University and Texas A&M) and the financial support from COLCIENCIAS accelerated the biotechnology development of these four organizations.
- The private sector focuses on the near-market and low-technology end of biotechnology and on horticultural crops, such as ornamentals and banana. These are high-value crops with a faster payback time. The most prominent private companies are Americaflor, GSL and ECB. The first two have recently been taken over by multinationals, to develop vertical control and integration of crops.
- The financial support from COLCIENCIAS between 1993 and 1996 was significant for the development of the biotechnology infrastructure, equipment and training programmes. Although agricultural biotechnology research expenditures grew annually, the percentage of agricultural biotechnology research to the total agricultural research expenditures in Colombia averaged 2% during 1985–1997.
- In 1997, there was a drastic reduction in COLCIENCIAS funding because of the financial crisis in Colombia. This caused bottlenecks in the provision of operational resources to projects that were approved before the crisis. It is critical to implement corrective measures to sustain the biotechnology capacity that some research organizations have generated.
- During 1985–1997, the number of researchers had grown much more rapidly than the research expenditures. This led to a significant decline in expenditure per researcher (in real international dollars from $60,600 in 1985 to $27,200 in 1997). There is a need for more national and institutional commitment to raise funding for agricultural biotechnology research. The more resources the researcher has, the higher the likelihood that research results and technology products will be achieved.
- Most of the Colombian organizations have a low ratio of technical support to researcher, with only one technical support person per researcher.
- About 50% of the surveyed research organizations showed a low researcher–manager ratio, which reflects a probable overmanagement of staff conducting biotechnology activities. This should be analysed more carefully.

- The public sector accounted for 61% (public research enterprise 34% and universities 27%) of the total expenditures during 1985–1997. The participation of the private sector (both commercial and non-commercial) is around 39%. Policies and incentives should be developed to encourage private-sector investment and participation in agricultural biotechnology research and development and closer public–private cooperation.
- Government has contributed the highest share of overall funding for agricultural biotechnology research activities in Colombia, representing about 50% in 1997. Donor share of total agricultural biotechnology research expenditures was 13%. These contributions went mainly to CORPOICA, Biogenesis and PUJ, largely to cover infrastructure and operational costs.
- In 1997, 60% of the researchers involved in biotechnology focused on crop research and 40% on livestock.
- Progress has been made in areas of patent law, plant breeders' rights and biosafety. The purpose of these regulations is to provide incentives for external investors in biotechnology to invest in Colombia.
- Colombia has experienced a lively expansion in biotechnology research since 1985. Despite the recent efforts of COLCIENCIAS in preparing the Biotechnology Strategic Plan, there is still no biotechnology policy in place to integrate and consolidate the research efforts (public organizations, academic entities and the private sector). Given the scarcity of resources, clear biotechnology research priorities should be defined and supported by a sound incentive scheme.

## References

Caswell, M., Fuglie, K. and Kotz, C. (1994) *Agricultural Biotechnology: An Economic Perspective.* Agricultural Economic Report No. 687, ERS, US Department of Agriculture (USDA).

COLCIENCIAS (1993) *Tecnologías de la Vida para el Desarrollo: Bases para un Plan del Programa Nacional de Biotecnología.* Bogotá, Colombia.

COLCIENCIAS (1999) *Biotecnología: Plan Estratégico 1999–2004.* Bogotá, Colombia.

De Kathen, A. (1999) *Transgenic Crops in Developing Countries.* Federal Environmental Agency, Texte, Berlin.

Downer, R., Dumbroff, E., Glick, B., Pasternack, J. and Winter, K. (1990) *Guidelines for the Implementation and Introduction of Agrobiotechnology into Latin America and the Caribbean.* Instituto Interamericano de Cooperacion para la Agricultura, San Jose, Costa Rica.

Fuglie, K., Ballenger, N., Day, K., Kotz, C., Ollinger, M., Reilly, J., Vasavada, U. and Yee, J. (1996) *Agricultural Research and Development: Public and Private Investments under Alternative Markets and Institutions.* Agricultural Economic Report No. 735, ERS, USDA.

Hodson, E. and Aramendis, R. (eds) (1995) *Biotecnología: Legislación y Gestión para América Latina y el Caribe.* COLCIENCIAS, Bogotá, Colombia.

Hodson, E. and Aramendis, R. (1998) *Biotecnología en Colombia. Programa Nacional de Biotecnología.* COLCIENCIAS, Bogotá, Colombia.

Jones, K.A. (1990) Classifying Biotechnologies. In: Persley, G.J. (ed.) *Agricultural Biotechnology: Opportunities for International Development*. CAB International, Wallingford, UK, pp. 25–28.

Rueda, M. (1996) Políticas, Prioridades y Programas que afectan a la Biotecnología Agrícola en Colombia. Country Profile prepared for the Seminar Transformación de las Prioridades en Programas Viables y Política Biotecnólogica Agrícola para América Latina, 6–10 October 1996, Peru (mimeo).

Salazar, S. (2001) *The Use of Proprietary Biotechnology Research Inputs at Selected Latin American NAROs*. International Service for National Agricultural Research (ISNAR), The Hague.

Sanint, L.R. (1995) *Crop Biotechnology and Sustainability: a Case Study of Colombia*. OECD Development Centre, Paris.

Sasson, A. (1993) *Biotechnologies in Developing Countries: Present and Future*, Vol. 1: Regional and National Surveys. UNESCO Future-Oriented Studies, Paris, France.

Spijkers, P. (1998) Aplicaciones biotécnologicos para agricultores de pequeña escala: el programa de DGIS en Colombia. In: Komen, J., Falconi, C. and Hernández, H. (eds) *Transformación de las Prioridades en Programas Viables: Actas del Seminario de Política Biotecnólogica Agrícola para América Latina*. Intermediary Biotechnology Service, ISNAR, The Hague.

Torres, R. (1993) *The Situation of Agricultural Biotechnology in Colombia*. ISNAR, The Hague.

Torres, R. and Falconi, C. (2000) *Agricultural Biotechnology Research Indicators: Colombia*. Discussion Paper No. 5, ISNAR, The Hague, 47 pp.

# Costa Rica

13

**Ana Sittenfeld, Ana Mercedes Espinoza, Miguel Munoz and Alejandro Zamora**

| | |
|---|---|
| Area ($km^2$) | 51,100 |
| Cropland | 6% |
| Irrigated cropland | 1200 $km^2$ |
| Permanent pasture | 46% |
| Population (1999 est.) | 3.67m |
| Population per $km^2$ | 68 |
| Annual population growth (1999 est.) | 1.89% |
| Life expectancy (men) | 73 yrs |
| (women) | 78 yrs |
| Adult literacy | 95% |
| GDP (1997 est.) | US$9 bn |
| GDP per head (1998 est.) | US$6700 |
| Growth in real GDP (1998 est.) | 5.5% |
| Agriculture as % of GDP | 18 |
| Agriculture products as % of total exports | 70 |
| Agricultural exports (US$) (1997) | 1.7 bn |

Major export commodities:
vegetables, fruit, ornamentals, coffee, banana, sugar

Major commodities:
coffee, banana, sugar, maize, rice, bean, potato, beef

## Summary

Research in agriculture in Costa Rica has generated a number of useful technologies, contributed to national food security and developed successful research systems on a few selected export crops, such as coffee and banana. Agricultural research has attempted to maximize production using high inputs that caused pollution and contamination of land, water and animal life.

Costa Rica is a small country that

has enjoyed a long history of conservation. The accelerated growth of protected lands, coupled with deforestation and lack of institutional coordination, led to the formulation of a National System of Conservation Areas (SINAC) in 1986. Today SINAC comprises a system of clearly defined protected areas encompassing about 25% of the national territory (1.2 million ha in 1997). Conservation areas are the main attraction for tourism, an industry that generated US$719 million in 1997.

Having a quarter of its territory separated for wild-land protection and realizing that only 15% of the soils are adequate for agriculture, Costa Rica needs to find ways to take advantage of its biodiversity. It is a major challenge for sustainable development to find innovative ways to link conservation and biotechnology to increase agricultural production on less land, with lower pesticide use, and to maximize the benefits of bioprospecting.

## Introduction

The initial commercial goals of agricultural biotechnology were directed at the markets of the industrial world. There is a growing realization, however, that agricultural biotechnology could make a valuable contribution towards solving the urgent problem of food supply, protecting the environment and reducing poverty in developing countries. There is no doubt that agricultural biotechnology has opened up new possibilities, particularly in crop and livestock development. The question remains as to just how agricultural biotechnology will increase the standard of living in developing countries. A sustainable strategy to provide food security for a growing population must promote biodiversity conservation and avoid further habitat loss of natural ecosystems. The strategy must also seek to: reduce unsustainable technologies, such as the overuse of chemical fertilizers and pesticides, unsustainable irrigation procedures and soil preparation methods that promote soil erosion; increase nutritional composition; reduce postharvest storage losses; and increase production from the present 2 billion t per year to 4 billion t. The strategy must also deal with issues of ethics, biosafety and intellectual property rights (IPR).

Food security today must be defined in terms of grains, meat and milk production and supply. Over the next 20 years there will be an increase in demand for meat, with most of the increased demand coming from developing countries, thus making investment in livestock research a necessity. The relative importance of different livestock species varies: cattle are generally more important for Latin America and the Caribbean, small ruminants in sub-Saharan Africa, small ruminants and buffalo in South Asia and pigs and poultry in East Asia and Latin America (Delgado *et al.*, 1999).

Strong science and technology development is fundamental to the successful use of biotechnology in agriculture. Developing countries are not homogeneous, however, in terms of scientific capabilities, social structures and economic goals, so there is no single solution for all countries. There are countries with no or very little capacity in agricultural biotechnology. They require different strategies from countries with an up-to-date biotechnology programme. The latter countries also have a national policy and strong connections within the country between the public and private sectors and between both those sectors and their equivalent sectors in more advanced countries (Brink *et al.*, 1999). It is not surprising, therefore, to find first-class biotechnology laboratories in China, India, Thailand, Brazil, Argentina, Mexico, Egypt and South Africa, which are perfectly capable of competing in the world of agricultural biotechnology.

In a recent study, Solleiro and Castañón (1999) indicate that Latin America has arrived late on markets for biotechnology products and services, a situation related to an industrial structure that is traditionally reluctant to introduce changes and has little capacity for R & D. The authors give several explanations: most R & D is conducted at universities and public-sector institutions, with minor participation of the industrial sector; human resources are not sufficient to cover all demands and biotechnology is not structured in a multidisciplinary way with capacity in molecular biology as well as management and marketing. A few successful efforts have incorporated competitive strategies combining in-house skills with excellent capabilities to locate, acquire and assimilate external technologies.

A significant portion of the improvements in agricultural biotechnology are being developed by and/or controlled by a few major multinational companies, making it more difficult for developing countries to access know-how in this area. For many of the food production companies the objective is to become more integrated by promoting vertical coordination of food systems, from the field to the supermarket. Different forms of acquisitions, mergers and alliances continue to be a dominant characteristic of the biotechnology industry (Thayer, 1998). By 1998, companies that were primarily chemical in origin had taken over most of the seed business. Through joint ventures and acquisitions, Monsanto and DuPont now market over half of the seeds for the two largest US crops, soybean and maize (Thayer, 1999).

Patenting and IPR are promoting privatization of scientific research in agricultural biotechnology and might increase the gap of biotechnology know-how and its applications between developing and industrial countries (Serageldin, 1999). Alternatives for developing countries include increasing research capacities at national institutions and at the international agricultural research centres (IARCs). This will allow the development of their own tools and know-how, which could be protected by IPR and to acquire the necessary negotiating power to exchange licences and implement strategic alliances with the private sector. The provision by donors of free licences, acquired from the private sector and academic institutions, for basic enabling biotechnologies should also be considered when supporting projects in developing countries and IARCs. Obtaining free licences from the private sector, to be applied in providing solutions to tropical crops that are not on the agenda of industries, offers another alternative for developing countries. In general, learning how to use IPR as a tool to advance biotechnology in developing countries, together with public and private investment, as well as new and imaginative public–private collaboration, is needed to promote technology transfer and better use of resources.

Some arrangements involving transfer of proprietary technologies by the private and public sectors in industrial nations, without royalties, to developing countries are already taking place (Krattiger, 1998). The benefit for developing countries is obvious; by increasing crop yields, genetically modified (GM) crops reduce the constant need to clear more land for producing food; seeds designed to resist drought and pests would be especially useful in tropical countries, where crop losses are often severe. Scientists in industrial countries are already working with colleagues and individuals in developing countries to increase yields of staple foods, to improve quality for better market acceptance and to diversify economies by creating new exports (Simon Moffat, 1999).

## Opportunities

Biodiversity-rich countries can take advantage of their biological/genetic resources from wild-land diversity, locally adapted varieties and races and wild relatives of crops to increase yields. This can be performed by applying agricultural biotechnology tools, by implementing bioprospecting activities and by establishing partnerships with public- and private-sector institutions in industrial and developing countries, including the IARCs. Investments in infrastructure are much lower than in any other high-technology field, with the exception of software development.

Several countries and institutions are implementing bioprospecting agreements with the private and public sector, based on the opportunities and obligations offered by the Convention on Biological Diversity and on the new developments in biotechnology and molecular biology, which are rapidly generating new tools and bioproducts. Bioprospecting collaborations are occurring in both developing and industrial countries (Sittenfeld, 1996; Varley and Scott, 1998). In this process, the definition of policies on access to genetic resources by governments and nations, as part of well-planned bioprospecting frameworks, are of particular importance for the success of national programmes. These activities integrate the search for compounds, genes and other nature-derived products with the sustainable use of biological resources and their conservation, along with scientific and socio-economic development of source countries and local communities.

Agricultural biotechnology, specifically the search for new genes for plant improvement, offers advantages to biodiversity-rich countries, compared with pharmaceutical research. Infrastructure and capital equipment costs are higher for the pharmaceutical area than for agricultural research (Tamayo *et al.*, 1997). The need for alternatives to production and protection of crops and livestock and the increasing capacity in biotechnology (e.g. differential gene expression techniques and genetic engineering) offer new opportunities for bioprospecting. Biotechnology can facilitate the transfer of several traits from wild biodiversity into cultivated crops. However, as with traditional plant breeding, there is a need to select the precise traits that consumers would reward in the market (Carter, 1996). Advances in biotechnology also provide choices of diversity beyond traditional use of *ex situ* collections in germ-plasm banks. It is important to incorporate *in situ* collections (in the form of wild biodiversity) into agricultural research. Together with this concept, the need to develop innovative systems to connect to agricultural practices, biodiversity conservation and intelligent use of biological resources becomes apparent (Sittenfeld and Lovejoy, 1996; Sittenfeld, 1998).

Many of the advances in agricultural biotechnology are developed in industrial countries, in close proximity to growing biotechnology companies, and therefore favour the agricultural practices of the industrial countries. This may pose a problem for the primarily agricultural economies of several countries in Latin America and other developing countries, because these developments may displace or transfer the production of these countries to the farm fields of the industrial countries, or even possibly to industrial bioreactors (Tamayo *et al.*, 1997). The concept of modern biodiversity prospecting, already proved in drug research (Sittenfeld and Villers, 1993, 1994), offers an alternative to this threat by transferring biotechnology to developing countries in exchange for access to their biological resources. This will enable developing countries to use their own biological resources while retaining a competitive edge with industrial countries. We can find examples of this practice in Mexico, Surinam, Peru, Argentina, Chile and Costa Rica.

The Instituto Nacional de Biodiversidad (INBio) in Costa Rica is negotiating agreements with scientific research centres, universities and private enterprise that are mutually beneficial to all parties (Sittenfeld, 1998). These pioneering agreements provide significant returns for Costa Rica while simultaneously assigning an economic value to natural resources and providing a new source of income to support biotechnology and the maintenance and development of the country's conservation areas.

Linking biotechnology and biodiversity through modern bioprospecting requires the creation and implementation of adequate frameworks integrating favourable macropolicies, biodiversity inventories and information systems, technology access and business development (see Platais and Persley, 2001). The principle of bioprospecting may be simple, but the link between biotechnology, biodiversity conservation and its sustainable use requires several considerations, including: a realization that a wider range of skills are required for research, product development and approval; the creation, use and management of multidisciplinary teams dealing with the complexities of legal and regulatory frameworks for biotechnology and biodiversity conservation and use; and the use of advanced applications of biotechnology to broader arrays of bioresources. Finally, understanding the opportunities and problems derived from international collaborative research and the linkages with commercial organizations represents a key point for favourable bioprospecting activities (Sittenfeld *et al.*, 1999).

## Linking biodiversity and biotechnology

Agriculture has been one of the most important sectors for the economy of Costa Rica, promoting democracy, national values and political stability. Agricultural expansion, however, has resulted in poor natural resource management, with low value-added prices for most of the crops (Mateo, 1996). The agricultural sector, although still contributing about 18% of the gross national product (GNP) and representing 70% of the total exports from 1970 to 1997 (Proyecto Estado de la Nación, 1998), is currently undergoing changes caused by shifts and pressures of globalization and fluctuating export prices in coffee and banana. A few successful exceptions are niche export markets for non-traditional products, such as high value-added vegetables, fruits and ornamentals (Mateo, 1996). In 1997, agricultural exports accounted for US$1.7 billion, although the size of the crop area diminished by 32%, from 179,034 ha in 1970 to 120,118 ha in 1997. The active population in the agricultural sector dropped from 25.3% in 1990 to 20.2% in 1997.

### *Rice Biotechnology Programme at the Centre for Research in Cellular and Molecular Biology (CIBCM), University of Costa Rica*

Rice is the most important staple crop in Costa Rica, providing almost one-third of the daily caloric intake, with a per capita consumption of 55 kg year$^{-1}$. Production is based on rain-fed and irrigated rice varieties developed several decades ago at the Centro Internacional de Agricultura Tropical (Cali, Colombia). Due to a narrow genetic background, however, all the varieties are susceptible to similar pests and diseases, such as planthoppers, rice hoja blanca virus (RHBV) and rice blast fungus *Magnaporthe grisea*, as well as physiological disorders, such as iron toxicity and zinc deficiency.

Because of a lack of resistance or tolerance to these factors, the use of pesticides and fungicides has increased costs, which reduces the profit margins and competitiveness of rice production in Costa Rica. Moreover, yield has remained fixed at 4.5 t ha$^{-1}$, leading to a strong dependency on international markets. A strategy based on pesticide spraying is also leading to pollution of water and wildlife refuges. Weed control, especially of red rice, a complex of *Oryza* species, represents nearly one-third of production costs.

The Rice Biotechnology Programme has been supported by several institutions, including the Rockefeller Foundation and the Costa Rican–United States Foundation for Cooperation (CRUSA). It is centred on the use of biotechnology to make biodiversity available for crop improvement and to diminish or eliminate some constraints on rice production in Costa Rica. The strategy includes the molecular characterization of wild rice germ-plasm found in the country, which may harbour useful agronomic traits for future use in crop improvement. A second approach is bioprospecting for bacterial genes with insecticide activity isolated from different genera, such as *Bacillus thuringiensis*, *Photorhabdus* spp. and *Xenorhabdus* spp., in different ecosystems. Isolated genes might be incorporated into the rice genome through genetic engineering. The strategy also includes genetic characterization of *M. grisea* lineages, in both cultivated and wild rice species, to define sources of disease resistance.

Facilities were developed for plant genetic engineering at the CIBCM to offer a new tool for rice breeding programmes. The first attempt at genetic transformation of rice was focused on the development of commercial rice cultivars resistant to hoja blanca, using viral genes and modified versions of those genes, which upon expression in plants may induce tolerance or resistance to the disease. This project started in 1989, with the molecular characterization and sequencing of the RHBV, the development of plant tissue-culture protocols for regeneration of Costa Rican 'indica' rice varieties and epidemiological studies on transmission and dispersion of RHBV by its insect vector, the planthopper *Tagosodes orizicolus*, which is also a pest of rice. Transgenic plants were produced using the RHBV coat-protein gene, as well as modified versions of the gene.

The population of Costa Rica is increasing and cultivated land area is diminishing, so our ultimate goal is to increase yield per area through the use of biotechnology. In the long term we expect to have a pool of useful genes from wild rice relatives, bacteria or even non-related plants and to transfer them to commercial rice cultivars. Wild rice species have proved to be useful resources for enriching the genetic pool of cultivated rice. Interspecific crosses with *Oryza rufipogon* have increased yield up to 20%. Also, the Xa21 gene from *Oryza longistaminata* has been cloned and introduced into the rice genome, thus conferring the plants with resistance to *Xanthomonas oryzae*. Some of the characteristics that could potentially be used for the improvement of rice are: pests and pathogen resistance, higher protein content, plant vigour and tolerance to high metal concentrations, salinity and soil acidity.

The Rice Biotechnology Programme includes research to identify, map and characterize the native relatives of rice that occur in Costa Rica, which is conducted at CIBCM in collaboration with the International Rice Research Institute (IRRI), in the Philippines. The location of the plants was recorded with a geographical positioning system (GPS) and the distribution of wild rices was correlated with a series of ecological and geographical variables. The identification and characterization of the wild species were done by morphological methods and the genetic variability of these species is being studied using rice microsatellites and isozymes.

Populations of three of the four *Oryza* species reported for tropical America have been found in natural ecosystems throughout the country, accounting for three of the six described genome types of *Oryza*. Of these, *Oryza latifolia* is the most variable, abundant and widely distributed. *Oryza grandiglumis* and *Oryza glumaepatula* are reported for the first time for Costa

Rica. These two species have restricted distributions and need to be preserved, since they are not appropriately protected at the moment. Furthermore, two native populations of putative sterile hybrids that reproduce asexually have been found.

## Constraints

We have two concerns. First, IPR on the vectors and genes used in the transformation processes may be under patent protection in the hands of private companies and academic institutions. This principle also applies to patents on technologies and tools used, such as plant transformation systems, selectable markers and gene expression technologies. When broad patents or patents on basic research are obtained by the private sector, the consequences for public research products are important. This is the case for biolistics (DuPont), *Agrobacterium*-mediated transformation (Japan Tobacco) and coat-protein-mediated resistance (Monsanto). If these technologies are used only for research purposes, there is a general agreement that no infringement occurs. However, this is not the case when research is translated into products in the market-place. This situation is affecting the development of transgenic plants in Costa Rica. In addition, the Patent Law of 1983 excludes the protection of biotechnology products and procedures. The law will be changed to meet Trade-Related Aspects of Intellectual Property Rights (TRIPs) requirements.

In Costa Rica, as in other countries, the distribution channels and agricultural extension in the public sector to reach farmers' fields are in a process of change towards privatization. Broad IPR protection of enabling technologies in the hands of the private sector might have a serious impact, as seed distribution channels are undergoing privatization around the world (Spillane, 1999).

The second concern is related to biosafety regulations and the risk of using modified rice cultivars in tropical environments. No documented experience on this topic has yet been published, to the best of our knowledge. It is one of the most contentious arguments against the use of transgenic plants, since it is assumed that wild relatives might, under specific conditions, hybridize and give rise to a new hybrid, which may pose a threat to agriculture if it behaves as a weed. No scientific evidence has been presented to support this argument, though a large body of speculation is shaping the opinion about plant biotechnology among consumers and even regulatory offices. The mapping of native relatives of rice species in Costa Rica is providing important information for selecting field-trial locations and crop areas. Conducting this type of study, in connection with the production of different transgenic rice plants, offers an interesting model to prevent risks associated with agricultural biotechnology developments. It is important to note that in 1997 Costa Rica included biosafety regulations in its Phytosanitary Protection Law No. 7664.

## Benefits

Will Costa Rica benefit from agricultural biotechnology? The answer may lie more heavily in the solution of the two concerns above than in the ingenuity of the plant breeders and molecular biologists.

INBio, established in 1989, is a private non-profit research institute. Its mission is to promote greater awareness of the value of biodiversity and thereby promote its conservation and improve the quality of life of Costa Rican society. The institute generates knowledge about biodiversity and disseminates and promotes the sustainable use of biological and genetic resources. Several of INBio's programmes, including its National Biodiversity Inventory, Bioprospecting, Information Management and Information Dissemination and Conservation Programme, document the biodiversity that exists in Costa Rica, where it can be found and how the country can find sustainable, non-damaging ways to use it and conserve it (Tamayo *et al.*, 1997). The collaborative agreement established between INBio and the Ministry of the Environment and Energy (MINAE) provides the framework for inventory and bioprospecting activities in collaboration with the SINAC. INBio, through specific access permits, collects samples for its Inventory and Biodiversity Prospecting Divisions and shares intellectual and monetary benefits with MINAE.

Bioprospecting involves the screening of biological and genetic resources for their potential use and the development of innovative strategies for capacity building, adding value and the generation of resources to invest in conservation activities. Within this framework bioprospecting is carried out in collaboration with local and international research centres, universities and the private sector. The set of criteria used by INBio to define its research agreements include access, equity and transfer of technology and training. Agreements stipulate that 10% of research budgets and 50% of any future royalties be awarded to MINAE for investment in conservation, according to the Biodiversity Law of 1998. The remainder of the research budget supports in-country capacity in biotechnology and value-added activities, also orientated to conservation and the sustainable use of biodiversity.

Management requirements for a successful bioprospecting enterprise, based on the INBio experience, include the following:

- Defining and implementing a bioprospecting framework, meaning favourable macropolicies, biodiversity inventories, information management systems, technology access and business development.
- Creating interdisciplinary and multidisciplinary teams of scientists, lawyers, conservation managers and business developers.
- Distributing the benefits obtained from bioproducts into building biotechnology capacity and improving biological resource management.

INBio builds on sound biodiversity knowledge, which helps to define market needs, major actors and national scientific and technological capacities

(Sittenfeld, 1996; Tamayo *et al.*, 1997). The principal markets for bioprospecting are the pharmaceutical, agricultural and biotechnological sectors, with an estimated market size of over US$600 billion worldwide (Sittenfeld *et al.*, 1999). Important requirements for bioprospecting include knowledge of national and institutional strengths and weaknesses, market surveys and evaluation of conservation needs. Most of INBio's bioprospecting activities are concentrated on the development of new pharmaceutical products; however, the basic issues and strategies can be applied to the agricultural sector as well.

Because 'raw' biological samples have low market value (Reid *et al.*, 1993), bioprospecting should seek to increase value by moving beyond simple resource collection and distribution services. Research contracts should concentrate on augmenting the value of biological resources by carrying out research in the source country. Additionally, involving in-country academics and researchers ensures that technologies transferred or accessed remain in the developing country. Increasing value is particularly important when negotiating royalty fees. In general, royalties for raw samples and collecting information are usually low, but adding information on the activity, structure and use of compounds and genes will allow increased sharing of profits, up to 15% or more, depending on the area of activity and market size of the product (Reid *et al.*, 1993; Ten Kate, 1995).

Guidelines provided by the Convention on Biological Diversity and research experiences with different commercial and academic entities allow INBio to follow basic rules, such as the fair and equitable sharing of benefits, the implementation of collection methods with reduced impact on biodiversity, technology transfer, biotechnology capacity building and up-front contribution to conservation activities. Examples of INBio agreements with academic and commercial entities are described elsewhere (Sittenfeld and Villers, 1993, 1994; Mateo, 1996; Nader and Rojas, 1996; Sittenfeld, 1996).

Recent experience in biodiversity prospecting negotiations have succeeded in establishing favourable terms for technology transfer, royalties and direct payments, among others things, for INBio and Costa Rica's conservation areas. Agreements have been developed between INBio and public and academic research institutions in Costa Rica and abroad, INBio and Merck & Co. Inc., INBio and the British Technology Group, INBio and Hacienda La Pacífica, INBio and the Bristol Myers Squibb Corporation, and others. The issue of benefits accrued from bioprospecting is difficult, given the inherent complexities of assigning value to the accumulated knowledge of biodiversity, the transfer of know-how and technology and enhanced capacity building. Up to now, products obtained from samples processed by INBio have not reached the marketplace. From 1992 to February 1998 INBio conducted bioprospecting agreements worth over US$6 million. The use of this money can be broken down as follows: US$3.5 million for investments and research expenses at INBio (taxonomy, information management and biotechnology), US$1.2 million that has been distributed to MINAE and the conservation areas; and US$0.8 million to support biotechnology development at public universities. It is important to take into consideration that the figure of over US$2.5 million for conservation and biotechnology development is significant for a country the size of Costa Rica, with a GNP of only US$9 billion for 1997 (Proyecto Estado de la Nación, 1998). MINAE has used its share to support the management and upkeep of Costa Rica's National Park at Coco Island, a unique site. This is a good example of direct bioprospecting benefits flowing to conservation (Mateo, 1996).

The economic value of bioprospecting should not be overestimated. Bioprospecting can complement other activities to advance human development and biodiversity conservation. Recent national policies, for example, established in Costa Rica to promote ecotourism, to protect wild-lands and to stimulate private reafforestation and secondary forests, together with the promotion of reafforestation programmes on carbon offset for carbon fixation, produced a forest coverage of 40%, with an increase of 2.6% of secondary forest in the last year. The deforestation rate went from 22,000 ha in 1990 to 8000 ha in 1994 and continues to decline (Proyecto Estado de la Nación, 1998).

## International Cooperation

Biotechnology is a necessary, but not sufficient, condition to advance social good in food and medicine. However, the IARCs, together with the World Bank and the regional development banks, have a pivotal role in ensuring that agricultural biotechnology has a positive impact on the standard of living, food security and poverty reduction in developing countries, by assisting these countries to take competitive effective advantage of their natural resources. Some considerations are:

- Establishment of a system of soft loans for biotechnology development and capacity building.
- Provide national agricultural research systems (NARS) with information, in particular with constant reviews of worldwide developments in the area of agricultural biotechnology and with studies of the likely impacts and appropriate developments in agricultural biotechnology for the country or the region.
- Identification of realistic objectives and strategies for the sustainable use of biodiversity and guidance in the implementation of adequate bioprospecting frameworks.
- Identification of opportunities to avoid or reduce the negative impacts of agricultural biotechnology.
- Promote activities to increase North/South, South/South and South/North interactions and the understanding of biosafety, biodiversity conservation and national capacity development in both science and markets.

- Development of in-house and in-country agricultural biotechnology research and negotiating skills, to enhance bargaining power when accessing IPR from the private sector, together with the provision by donors and IARCs of licences for enabling technologies, acquired from the private and academic sectors, to NARS.
- Promote cooperation between national and international public and private sectors, to increase food production and well-organized distribution systems in the country. In particular, more emphasis should be devoted to applications of biotechnology in livestock research.
- Increase public-sector biotechnology R & D of traits where the technology is not economically attractive to the private sector in the short term.
- Promote education in key areas for agricultural biotechnology development and public awareness on biotechnology.

## References

Brink, J.A., Prior, B. and DaSilva, E.J. (1999) Developing biotechnology around the world. *Nature Biotechnology* 17, 434–436.

Carter, M.H. (1996) *Monsanto Company: Licensing 21st Century Technology*. Case No. 596–034, Harvard Business School Publishing, Boston, Massachusetts.

Delgado, C., Rosegrant, M., Steinfeld, H., Ehui, G. and Courbois, C. (1999) Livestock in 2020: the next food revolution. In: IFPRI/FAO/ILRI (eds). *Food, Agriculture and the Environment*. Discussion Paper 28 (International Food Policy Research Institute/Food and Agriculture Organization/International Livestock Research Institute), Washington, DC.

Krattiger, A. (1998) *The Importance of Ag-biotech to Global Prosperity*. ISAAA Briefs No. 6, International Service for the Acquisition of Agri-Biotech Applications, Ithaca, New York.

Mateo, N. (1996) Wild biodiversity: the last frontier? The case of Costa Rica. In: Bonte-Friedheim, C. and Sheridan, K. (eds) *The Globalization of Science: the Place of Agricultural Research*. International Service for National Agricultural Research (ISNAR), The Hague, Netherlands, pp. 73–82.

Nader, W. and Rojas, M. (1996) Gene prospecting within the wild life: new incomes for biodiversity conservation through royalties. *Genetic Engineering News* 16, 35.

Platais, G.H. and Persley, G.J. (2001) *Biodiversity and Biotechnology: Contributions to and Consequences for Agriculture and the Environment*. Environment Department, World Bank, Washington, DC.

Proyecto Estado de la Nación (1998) *Estado de la Nación en Desarrollo Sostenible*, No. 4. Editorama S.A, San José, Costa Rica.

Reid, W.V., Laird, S., Meyer, C.A., Gámez, R., Sittenfeld, A., Janzen, D.H., Gollin, M.A. and Juma, C. (1993) *Biodiversity Prospecting: Using Genetic Resources for Sustainable Development*. World Resources Institute, Washington, DC.

Serageldin, I. (1999) Biotechnology and food security in the 21st century. *Science* 285, 387–389.

Simon Moffat, A. (1999) Crop engineering goes south. *Science* 285, 370–371.

Sittenfeld, A. (1996) Issues and strategies for bioprospecting. *Genetic Engineering and Biotechnology (UNIDO, Emerging Technology Series)* 4, 1–12.

Sittenfeld, A. (1998) Biotecnología, prospección de la biodiversidad y acceso a los recursos genéticos: cuestiones para América Latina. In: *Transformación de las prioridades en programas viables. Actas del Seminario de política biotecnológica agrícola para América Latina. Perú, 6 al 10 de Octubre de 1996.* La Haya/México, D.F. Intermediary Biotechnology Service/CamBio Tec. pp. 16–23.

Sittenfeld, A. and Lovejoy, A. (1996) Biodiversity prospecting frameworks: the INBio experience in Costa Rica. In: McNeely J.A. and Guruswamy L.D. (eds) *Their Seed Preserve: Strategies for Protecting Global Biodiversity*. Duke University Press, Durham and London. pp. 223–244.

Sittenfeld, A. and Villers, R. (1993) Exploring and preserving biodiversity in the tropics: the Costa Rican case. *Current Opinion in Biotechnology* 4 (3), 280–285.

Sittenfeld, A. and Villers, R. (1994) Costa Rica's INBio: collaborative biodiversity research agreements with the pharmaceutical industry. Case study 2. In: Meffe, G.K., Carroll, C.R. and contributors (eds) *Principles of Conservation Biology*, Sinauer Associates, Sunderland, Massachusetts, pp. 500–504.

Sittenfeld, A., Lovejoy, A. and Cohen, J. (1999) Managing bioprospecting and sustainable use of biological diversity. In: Cohen, J. (ed.) *Managing Agricultural Biotechnology – Addressing Research Programme Needs and Policy Implications for Developing Countries*. ISNAR, CAB International, Wallingford, UK, pp. 92–101.

Solleiro, J.L. and Castañón, R. (1999) Technological strategies of successful Latin American biotechnological firms. *Electronic Journal of Biotechnology* 2, 26–35.

Spillane, C. (1999) *Commission on Genetic Resources for Food and Agriculture. Recent Developments in Biotechnology as They Relate to Plant Genetic Resources for Food and Agriculture*. Background Study Paper No. 9, FAO (Food and Agriculture Organization), Rome, 64.

Tamayo, G., Nader, W.F. and Sittenfeld, A. (1997) Biodiversity for the bioindustries. In: Ford-Lloyd, B.V., Newbury, H.J. and Callow J.A. (eds) *Biotechnology and Plant Genetic Resources: Conservation and Use*. CAB International, Wallingford, UK, pp. 255–279.

Ten Kate, K. (1995) *Biopiracy or Green Petroleum? Expectations and Best Practice in Bioprospecting*. Overseas Development Administration (ODA), London.

Thayer, A.M. (1998) Living and loving life sciences. *Chemical and Engineering News* 76, 17–24.

Thayer, A.M. (1999) Transforming agriculture. *Chemical and Engineering News* 77, 21–35.

Varley, J.D. and Scott, P.T. (1998) Conservation of microbial diversity a Yellowstone priority. *ASM News* 64, 147–151.

# Index

Figures in **bold** indicate major references.
Figures in *italic* refer to diagrams, photographs and tables.

Browse Read and Buy
www.cabi.org/bookshop
ANIMAL & VETERINARY SCIENCES
BIODIVERSITY CROP PROTECTION
HUMAN HEALTH NATURAL RESOURCES
ENVIRONMENT PLANT SCIENCES
SOCIAL SCIENCES
Subjects
Search
Reading Room
Bargains
New Titles
Forthcoming
CABI Publishing
A division of CAB International
Online BOOK SHOP
Crop Pollination by Bees
Keith S. Delaplane and Daniel F. Mayer
A DICTIONARY OF Entomology
G Gordh and D H Headrick
Order & Pay Online!
Principles of CATTLE PRODUCTION
C.J.C. Phillips
Seeds
THE ECOLOGY OF REGENERATION IN PLANT COMMUNITIES
2nd EDITION
Edited by Michael Fenner
FULL DESCRIPTION
BUY THIS BOOK
BOOK OF THE MONTH
Tel: +44 (0)1491 832111 Fax: +44 (0)1491 829292